How to Achieve ISO 9000 Registration Economically and Efficiently

QUALITY AND RELIABILITY

A Series Edited by

EDWARD G. SCHILLING
Coordinating Editor
Center for Quality and Applied Statistics
Rochester Institute of Technology
Rochester, New York

RICHARD S. BINGHAM, JR.
Associate Editor for
Quality Management
Consultant
Brooksville, Florida

LARRY RABINOWITZ
Associate Editor for
Statistical Methods
College of William and Mary
Williamsburg, Virginia

THOMAS WITT
Associate Editor for
Statistical Quality Control
Rochester Institute of Technology
Rochester, New York

ADDITIONAL VOLUMES IN PREPARATION

How to Achieve
ISO 9000
Registration
Economically
and
Efficiently

Gurmeet Naroola, P.E.
Santa Clara, California

Robert Mac Connell, P.E.
Sunnyvale, California

Marcel Dekker, Inc.   New York•Basel•Hong Kong

Library of Congress Cataloging-in-Publication Data

Naroola, Gurmeet.
How to achieve ISO 9000 registration economically and efficiently / Gurmeet Naroola, Robert Mac Connell.
p. cm. — (Quality and reliability; 48)
ISBN 0-8247-9758-2 (alk. paper)
1. ISO 9000 Series Standards. I. Mac Connell, Robert. II. Title.
III. Series.
TS156.6.N37 1996
658.5'62—dc20

96-13968
CIP

The publisher offers discounts on this book when ordered in bulk quantities. For more information, write to Special Sales/Professional Marketing at the address below.

This book is printed on acid-free paper.

MARCEL DEKKER, INC.
270 Madison Avenue, New York, New York 10016

Current printing (last digit):
10 9 8 7 6 5 4 3 2 1

PRINTED IN THE UNITED STATES OF AMERICA

*To my family (Mom, Dad, Bua, Mini, Sonu),
my teachers (John, ToshiSan, KenjiSan),
my friends, and someone special in my
life. We are going places.*

Gurmeet Naroola

*To Claris, my wife of forty years, who is the
real quality in my life.*

Bob Mac Connell

About the Series

The genesis of modern methods of quality and reliability will be found in a simple memo dated May 16, 1924, in which Walter A. Shewhart proposed the control chart for the analysis of inspection data. This led to a broadening of the concept of inspection from emphasis on detection and correction of defective material to control of quality through analysis and prevention of quality problems. Subsequent concern for product performance in the hands of the user stimulated development of the systems and techniques of reliability. Emphasis on the consumer as the ultimate judge of quality serves as the catalyst to bring about the integration of the methodology of quality with that of reliability. Thus, the innovations that came out of the control chart spawned a philosophy of control of quality and reliability that has come to include not only the methodology of the statistical sciences and engineering, but also the use of appropriate management methods together with various motivational procedures in a concerted effort dedicated to quality improvement.

This series is intended to provide a vehicle to foster interaction of the

elements of the modern approach to quality, including statistical applications, quality and reliability engineering, management, and motivational aspects. It is a forum in which the subject matter of these various
areas can be brought together to allow for effective integration of
appropriate techniques. This will promote the true benefit of each,
which can be achieved only through their interaction. In this sense, the
whole of quality and reliability is greater than the sum of its parts,
as each element augments the others.

The contributors to this series have been encouraged to discuss
fundamental concepts as well as methodology, technology, and procedures at the leading edge of the discipline. Thus, new concepts are
placed in proper perspective in these evolving disciplines. The series is
intended for those in manufacturing, engineering, and marketing and
management, as well as the consuming public, all of whom have an
interest and stake in the improvement and maintenance of quality and
reliability in the products and services that are the lifeblood of the
economic system.

The modern approach to quality and reliability concerns excellence:
excellence when the product is designed, excellence when the product is
made, excellence as the product is used, and excellence throughout its
lifetime. But excellence does not result without effort, and products and
services of superior quality and reliability require an appropriate combination of statistical, engineering, management, and motivational effort.
This effort can be directed for maximum benefit only in light of timely
knowledge of approaches and methods that have been developed and
are available in these areas of expertise. Within the volumes of this
series, the reader will find the means to create, control, correct, and
improve quality and reliability in ways that are cost effective, that
enhance productivity, and that create a motivational atmosphere that is
harmonious and constructive. It is dedicated to that end and to the
readers whose study of quality and reliability will lead to greater understanding of their products, their processes, their workplaces, and themselves.

Edward G. Schilling

Preface

ISO 9000 has become the de facto Quality System Standard and companies worldwide are facing the need to become registered. There is confusion and misconceptions about what has to be done, and some people think that registration is costly and time consuming and does not produce substantial results.

We disagree with these attitudes!

This book provides a unique hands-on, step-by-step method by which to achieve ISO 9000 registration efficiently and economically. It carefully models a number of successful registration efforts personally conducted by the authors using the unique TAP-PDSA approach.

This is an extremely user-friendly book. The chapters are organized sequentially to take the reader from the beginning of a registration effort to "life after registration."

Chapter 1 introduces the user to a unique and proven method (TAP-PDSA) to achieve registration. *Chapter 2* discusses the "must know" ISO 9000 knowledge required before starting a registration effort. *Chapter 3* establishes the framework required to launch a successful registration effort. It stresses that the primary goals of a registration are to improve the quality system in order to achieve customer satisfaction and profitability. *Chapter 4* covers the first few important activities required to "jump start" the registration project. *Chapter 5* explains in detail each of the TAP elements as they relate to an ISO 9000 registration effort. A detailed ISO 9000 "TAP-PDSA" registration plan is included. *Chapter 6* addresses the functions of a registrar, the importance of early registrar selection, the criteria, and process of selection. *Chapter 7* explains the measurements in relation to the project and the quality system. *Chapter 8* shows to the reader the quality system design and documentation processes. *Chapter 9* discusses life after registration.

The primary intended audiences are: ISO Project Leaders, ISO Project Team, Quality Management, Manufacturing Management, and Materials Management. This book can also be used by consultants, educators, trainers, institutions, and professional societies.

We hope you will find the information contained here of importance. Contact us and let us know what you think. Your suggestions and comments are very welcome.

Gurmeet Naroola
Robert Mac Connell

Contents

List of Figures

List of Tables

List of Abbreviations

4 M's:	Man, Machine, Material, and Method
ABC:	Activity-Based Costing
AFNOR:	Association Francaise de Normalisation
ANSI:	American National Standards Institute
AQL:	Average Outgoing Quality Level
ASQC:	American Society of Quality Control
BSI:	British Standards Institute
Cpk:	Process Capability Indices
DIS:	Draft International Standard
E:	Environment
EN:	European Norm
EU:	European Union
IEC:	International Electrotechnical Commission
JIT:	Just-In-Time
MIL-Q:	Military Quality
MIS:	Management Information Systems
MOU:	Memorandum of Understanding
MRP:	Material Requirement Planning

NAA:	Needs Assessment Audit
NACCB:	National Accreditation Bodies
NNI:	Netherlands Normalistatie Instituut
PDSA:	Plan, Do, Study, Act
PIP:	Preliminary ISO Registration Plan
QA:	Quality Assurance
QCD:	Quality, Cost, and Delivery
QSD:	Quality System Documentation
RAB:	Registrar Accreditation Board
RvC:	RaadVoor de Certificate
SC:	Sub-Committee
SCC:	Standards Council of Canada
SIC:	Standard Industry Code
TAG:	Technical Action Group
TAG:	Technical Advisory Group
TAP:	Train, Audit, and Plan
TC:	Technical Committee

How to Achieve ISO 9000 Registration Economically and Efficiently

Chapter 1
TAP: The Philosophy

As with almost everything, the world turns, cycle times get shorter and shorter, and competition is becoming more fierce. To survive in these increasingly difficult times, quality products must be developed and delivered ahead of the competition.

Managers feel like they are on a treadmill with an alligator snapping behind as the treadmill goes ever faster and faster. No longer can a company take time to develop a system or plan by casually feeling its way through when the development time is so limited.

The point is that most people do not plan in an efficient manner. They adjust as they go along, but this process of continually adjusting and going backward can be wasteful. For this reason, it is important that the process of planning be carefully undertaken so that the outcome can be predicted with a high degree of *assurance, success, and confidence.*

This planning process is a critical step towards the realization of a goal. Good planning requires that an honest effort be devoted to training and education in researching the task prior to beginning the project. This process is spelled out in the following section as the "TAP" cycle for establishing a sound plan. The next phase, that of successful implementation of the plan, also involves a sequence of steps. This is described as the "Deming" or "PDSA" cycle for implementing the plan. It is interesting to note that **P**(Plan) remains as a common element to both the TAP and PDSA cycles. Details concerning both cycles presented here help achieve successful results in a timely, efficient, and therefore less costly manner.

1   Planning Cycle: TAP

The TAP cycle (Figure 1.1) has three essential elements:

- Train

- Analyze

- Plan

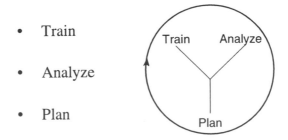

Figure 1.1   TAP

The TAP cycle requires that the group prepare itself through **training and education** to better predict future requirements, then perform an **analysis** of the existing situation to get a better understanding before developing an overall **plan** which has a high probability of success. Finally, a plan is detailed so that all will understand and know what is expected for implementation.

2   Implementation Cycle: PDSA

The implementation cycle (Figure 1.2) is shown as the Deming
PDSA cycle.

- Plan

- Do

- Study

- Act

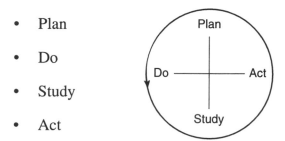

Figure 1.2   Deming PDSA

This cycle explains that the **plan** is implemented in the sequence of
first **do**ing what is detailed by the plan, then **study**ing the
effectiveness of the plan, and finally **act**ing on the recommended
changes to the plan. Thus the steps: Plan, Do, Study, Act.

3   Continuous Improvement Cycle: TAP-PDSA

The two cycles (TAP and PDSA) are illustrated in Figure 1.3. The
top cycle is the "driver" and results in proper planning. The bottom
cycle is the "driven" and this cycle implements the plan. The
diagram also shows plan as a common element to both cycles. It
also illustrates that these are never-ending cycles continually
revolving and driving the results higher and higher.

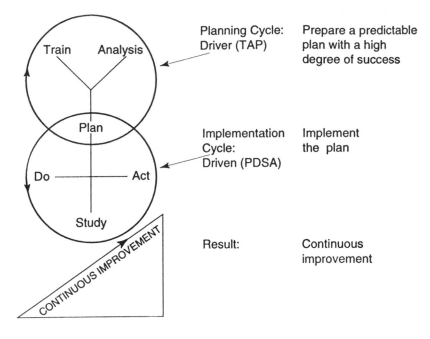

Figure 1.3   TAP-PDSA

3.1 Train

Training is the first step in the TAP cycle. It involves the acquisition of knowledge regarding project matters so that a plan that has a higher probability of success can be predicted. Train to understand the project: What is the purpose and objective? What is the scope of the project? and What are the requirements (cost, time, etc.)?

There are many additional benefits of training. Everyone learns to speak the same language and focuses in the same direction, creating a sound starting platform. Training and education also creates a body of highly trained employees who are the most important resources of the organization.

3.2 Analyze

This step not only provides the status quo, but also presents a clearer picture of the goals the group wishes to accomplish. Perform an analysis of the existing situation and identify current strengths and weaknesses. Imagine the future requirements. What direction are we headed towards in the future?

3.3 Plan

The better the training, the clearer the analysis, the better the plan. The planning stage may start with a choice between several suggestions. Which direction should we go? Who leads the project? Compare the possible outcomes of the possible choices. Of the several suggestions, which one appears to be most promising in terms of quality, cost, and delivery? The goal is to use the plan that has the highest probability of success. This plan involves a detailed roadmap to get from the present condition to the target.

3.4 Do

The do step implements the plan developed by the planning cycle.

3.5 Study

Part of the plan should provide measurement criteria to continuously monitor the project progress. This activity is performed at the study stage. Compare the results with the plan. Are we progressing according to the plan? Are the results acceptable? Does the plan require any modification?

3.6 Act

The outcome of the study stage dictates this step. If all is well, the implementation moves on. If the outcome did not provide the desired results, appropriate changes are made to the plan and the cycles are repeated. This leads back to the plan, which is continuously improved through ongoing training and analysis. This is like one gear driving another gear with TAP as the driver gear and PDSA as the driven gear.

4   TAP-PDSA and ISO 9000

TAP can be used in any project successfully to produce outstanding results. The ISO 9000 registration effort is no exception.

The ISO 9000 registration project kicks off with **training and education,** making sure that all involved talk the same language, that of ISO, and look in the same direction; again that of ISO registration.

Upon having a clear understanding of ISO, a detailed **analysis** of the existing quality system is performed against the ISO 9000 requirements. This step helps understand the status quo, weaknesses, and strengths of the quality system. It also helps to properly determine the resources needed (i.e., cost, manpower, etc.) to successfully achieve the target of registration. This analysis usually takes place in the form of the Needs Assessment Audit.

With a clear understanding of the present situation, target, and the resources required, an implementation **plan,** with a higher success rate and minimum revisions, can be predicted.

This plan is then implemented through the Do, Study, and Act steps of the Deming Cycle.

The benefits of this approach are many. The principal ones are:

*   A superior quality system will be developed

*   Registration will be achieved sooner

*   Registration costs will be significantly lower

Note: See the plan section in Chapter 5 for the ISO 9000 registration TAP-PDSA plan Gantt chart.

5   Conclusion

Together the TAP and PDSA cycles become very powerful tools for continuous improvement, with TAP as the driver or the planning cycle and PDSA as the driven or the implementation cycle.

Chapter 2
The Must Knows

This chapter answers in sufficient detail the "must know" questions most frequently asked prior to beginning an ISO 9000 registration project. This information provides the reader with a basic understanding regarding ISO, the ISO 9000 standards, and the registration process required to establish a strong foundation stone.

The questions and answers will be grouped by topic:

- ISO
- ISO/TC-176
- ISO 9000 Series of Standards
- Registrars
- Registration Process
- Guidance Standards

1   ISO

1.1   What Is ISO?

In the aftermath of World War II, when most of the industrial base in Europe had been badly damaged, a number of international conferences were held to facilitate industrial reconstruction with both allies and former enemies participating. One such conference's objective was to foster trade and commerce development among the countries emerging from the conflict through the development of uniform international manufacturing, trade, and communication standards. In 1946, the birth of the International Organization for Standardization (ISO) was the result of this conference. The founders felt that through the use of commonly accepted standards, countries would be better able to use each others commodities, manufactures, and products.

Today, ISO is a worldwide federation, headquartered in Geneva, Switzerland with over 100 member countries. It develops standards for all industries except those industries related to the electric and electronics disciplines, which are the venue of the International Electrotechnical Commission (IEC). Together, these two groups form the largest and most comprehensive world–wide, non–governmental forums for voluntary industrial and technical collaboration at the international level.

1.2   What Does ISO Mean?

While the ISO 9000 series continues to achieve greater recognition around the world, many executives, consultants, and journalists are still confused as to what the ISO name stands for. ISO doesn't stand for anything, although it functions as an acronym when referring to the Geneva-based International Organization for Standardization. According to ISO officials, the organization's short name was borrowed from the Greek word Isos, meaning "equal." Isos also is the root of the prefix "iso," which appears in "isometric" (of equal measure or dimension) and "isonomy" (equality of laws or of people before the law). Its election was based on the conceptual path taken from "equal" to "uniform" to "standard." In an attempt to make sense of the ISO name, many have misconstrued the organization's full name to be the International Standards Organization.

1.3 Who Were the Founding Members of ISO?

The founding members consisted of fourteen industrialized countries from Europe, the United States, and the British Commonwealth. The American National Standards Institute (ANSI) was the founding member representative for the United States.

1.4 Where Is ISO Located?

The headquarters secretariat for the International Organization for Standardization is located in Geneva, Switzerland.

Mailing address:

1, rue de Varmbe,
Caste postale 56,
CH1211 Geneve/Suisse.

Phone: 41 22 749 0111
Telex: 41 22 733 3430
Internet: CENTRAL@ISOCS.ISO.CH

1.5 How Does ISO Operate?

ISO work is decentralized, carried out by over 180 active technical committees (TCs) and over 620 active sub–committees (SCs). They are supported by technical secretariats in thirty-four countries. The General Secretariat in Geneva assists in coordinating ISO operations world-wide. Over 30,000 specialists in their fields develop international standards. These specialists are nominated by ISO members to participate in committee meetings, and to represent the consolidated views and interests of industry, government, labor, consumers, and other interested parties in the standards development process. The actual standards preparation is done by ISO Technical Committees (ISO/TCs) assembled and assigned responsibility for developing related standards. The working technical committees and sub–committees meet periodically in world-wide locations to develop or review standards work. TCs are assisted by member country Technical Action Groups (TAG) to provide counsel for standards under development or revision. Requests to develop new standards come from member countries to ISO, which evaluates the need and benefit before a TC is assembled.

Work of the International Organization of Standardization is supported through the sales of standards to users world-wide. Authorized languages for ISO standards are English, French, and Russian. Member countries often issue ISO standards as country standards using the language of the issuing country.

2   ISO/TC–176

2.1   What Is ISO/TC–176?

Technical work in the International Organization for Standardization is handled by technical committees (TCs). ISO/TC-176 is the 176th ISO technical committee and was formed in 1979 to address the needs for standards in Quality Management and Quality Assurance. ISO/TC-176 completed the development of the ISO 9000 core series of standards in 1987.

ISO technical committees are numbered serially as they are established, starting with TC-1 (screw threads) in 1946 to TC-207 (environmental management) in 1993. A number is never repeated, nor is a technical committee assigned a new project when its work is done.

2.2 How Is TC-176 Organized?

TC-176 consists of a central committee and three sub-committees (SCs) to deal with specific areas as shown in Figure 2.1.

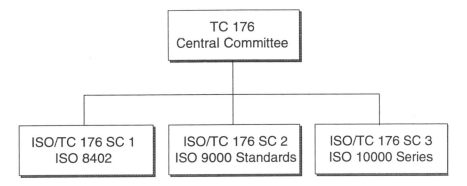

Figure 2.1   TC-176 Organization

ISO/TC-176 SC 1

SC 1, responsible for the development of concepts and terminology, created ISO 8402. The Secretariat is Association Francaise de Normalisation (AFNOR), France.

ISO/TC-176 SC 2

SC 2 was responsible for the development of quality systems guidelines and standards (ISO 9000 through 9004). Secretariat is British Standards Institute (BSI), United Kingdom.

ISO/TC-176 SC 3

SC 3 was responsible for the development of supporting technology guidelines (ISO 100XX series). Secretariat is Netherlands Normalisatie Instituut (NNI), Netherlands.

The five national associations participating in ISO/TC–176 as organizers are BSI, ANSI, SCC, AFNOR, and NNI.

2.3 What Is the Role of ISO/TC-176 in the United States?

The American Society for Quality Control (ASQC) chairs the U.S. Technical Advisory Group (TAG) to ISO/TC–176. The TAG is most visible as the group that takes the ISO 9000 series of documents and "Americanizes" them as the Q9000 series of standards. The U.S. TAG also works directly with the TC to draft, revise, and publish documents.

2.4 Where Can I Learn More About ISO/TC-176?

For a fee, interested individuals can obtain 'observer status' to ISO/TC–176. ISO publishes a newsletter and bulletin on the ISO 9000 activities including ISO/TC–176. The technical committee has published "Vision 2000, A Strategy for International Standards Implementation in the Quality Area During the 1990s." This publication contains valuable information concerning the long-term objectives of the technical committee.

3 ISO 9000 Series of Standards

3.1 What Are the ISO 9000 Series of Standards?

The ISO 9000 series are a set of generic standards that state the requirements for an acceptable quality management system. The standards are of two types:

• Conformance
  – ISO 9001
  – ISO 9002
  – ISO 9003

- Guidance

    - ISO 9000 is a guidance standard for selecting the proper conformance standard.

    - ISO 9004 is a guideline for quality management and quality system elements.

As the standards and guidelines have become more widespread, the guidelines have been expanded to cover more and more industry and service situations. For example, ISO 9000 has become ISO 9000–1, 9000–2, 9000–3, and 9000-4, and ISO 9004 has become 9004-1, 9004-2, 9004-3, 9004-4, 9004-5, and 9004-7. (See Section 6.2, "What Guidance Standards Are Available" for details.) The following diagram shows the structure of the ISO 9000 series of standards.

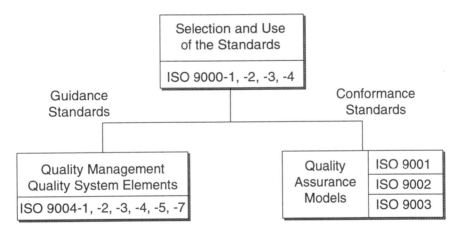

Figure 2.2   The Structure of ISO 9000 Standards

3.2 Why Were the ISO 9000 Standards Developed?

The initial need for the ISO 9000 International Quality System Standards was for two-party contractual situations, between customer and supplier. The objective was to increase customer confidence in the quality systems of its suppliers. With common standards, the approach also served the goal to make supplier quality systems more uniform. As manufacturers moved from vertical integration of manufacturing processes to increased dependence upon suppliers and sub–contractors, monitoring the quality of product through source inspections and supplier quality audits became more important.

The predecessors to the development of the International Quality Standard were national standards such as British Standard BS 5750 (a source document for the ISO 9000 series of quality standards), military standards (i.e., MIL–Q–9858A) in the United States, and a variety of industry standards and even company standards for inspection and quality assurance. As the influence of quality control and its recognition as a system accelerated in the last half of this century, more and more quality manuals included specifics for a good quality system (i.e., calibration, in–process inspection, and traceability). The ISO 9000 standards are unique standards for quality systems rather than specification or product standards.

Technical specifications may not, in themselves, guarantee that a customer's requirement will be consistently met. This led to the development of quality system standards, such as ISO 9000, and guidelines that complement relevant product or service requirements given in technical specifications.

3.3 What Is the Scope of the ISO 9000 Standards?

ISO 9000 standards are in use in every industrialized country and most emerging countries in the world. They are rapidly becoming the de facto global quality assurance standards for industry, business, education, and government. One of the achievements of the writers of the International Standards is that they are completely generic and can be applied to almost any situation where a quality system is needed.

3.4 What Industries Will Be Affected by ISO 9000?

Some products are already under the European Union (EU) requirements that the quality system of the manufacturing site conform to an ISO 9000 standard. It is certain that the list will grow and any industry that is "regulated" in some form will be targeted by the EU to comply with requirements for ISO 9000, or the equivalent national product quality standard in the near future. It is not unreasonable to assume that the requirement will become almost universal by the end of the decade.

To write authoritatively about the European Union direction and requirements is to be out–of–date tomorrow. For American industry, there exists great uncertainty surrounding EU directives and ISO 9000 registration. These uncertainties are creating mine fields for those companies wishing to export products to the EU countries. Directives have already been prepared with application to toys, construction products, pressure vessels, electromagnetic capability, machinery, personal protective equipment, gas appliances, non–automatic weighing instruments, medical devices, and telecommunications terminal equipment.

The key caveat is to be aware that a quality system must be certified by the "right" registrar, who is properly accredited to those who certify the product as being acceptable to an EU requirement. Again, as emphasized, ISO 9000 registration is only a component of a business strategy and always must be kept in that frame of reference.

3.5 How Do the Conformance Standards Relate to Each Other?

Of the three conformance standards, ISO 9001 is the most complete. It includes all elements in the production cycle, from design through servicing, and contains 20 requirements. ISO 9002 is the same as ISO 9001 except there are no requirements for design control. ISO 9003 is for final inspection and test only.

Table 2.1:   List of Quality System Requirements

Quality System Requirements		9001	9002	9003
1.	Management Responsibility	X	X	X
2.	Quality System	X	X	X
3.	Contract Review	X	X	X
4.	Design Control	X		
5.	Document and Data Control	X	X	X
6.	Purchasing	X	X	
7.	Control of Customer-Supplied Product	X	X	X
8.	Product Identification and Traceability	X	X	X
9.	Process Control	X	X	
10.	Inspection and Testing	X	X	X
11.	Control of Inspection, Measuring, and Test Equipment	X	X	X
12.	Inspection and Test Status	X	X	X
13.	Control of Non-Conforming Product	X	X	X
14.	Corrective and Preventive Action	X	X	X
15.	Handling, Storage, Packaging, Preservation, and Delivery	X	X	X
16.	Control of Quality Records	X	X	X
17.	Internal Quality Audits	X	X	X
18.	Training	X	X	X
19.	Servicing	X	X	
20.	Statistical Techniques	X	X	X

3.6 How Do I Select a Proper Standard for Use?

ISO 9000-1, Guidelines for the Selection and Use, was written to assist in selection of the proper standard to use. However, there is a very simple rule of thumb:

- ISO 9001 – If you design products

- ISO 9002 – If you manufacture only

- ISO 9003 – If you neither design nor manufacture

The best advice is to evaluate the three core standards, ISO 9001, ISO 9002, and ISO 9003, and make a confident selection.

3.7 How Are ISO 9000 Standards Interpreted?

Remember that the standards prescribe what is required for a compliant quality system, not how to meet the requirements. Fortunately, the standards state the requirements rather succinctly and the latest revisions (1994) were primarily for the purpose of removing inconsistencies, ambiguities, and unneeded vagueness. There is no official body that provides interpretations and this would be difficult, since no two applications are identical. The best interpreters are the personnel who have been well trained in the ISO 9000 International Standards.

The writers of the ISO International Standards prepared the text of the documents to carefully convey the meaning of the words used. No document in the quality manual should be prepared without first understanding the complete direction provided by the text for each clause of the International Standard.

Definitions for words not in common use are provided in the Vocabulary Standard: ISO 8402. Additional definitions may be found in the introduction clauses of the text itself. Other word definitions should be checked in any good dictionary. How they are used in the sentence structure is most important. Here are some examples:

Shall – Without question, "shall" is the most important word in the ISO 9000 Standards.

- "shall define and document"(4.1.1)
- "shall ensure" (4.1.1)
- "shall appoint"
- "shall review" (4.1.3), etc.

"Shall" is a command, "you will do it!" To avoid confusion, the word "will" is not used. Modifiers, such as ".....on occasion, shall" are not used. Shall tells the readers that something is required to be done and objective evidence of actions performed is needed. You must be able to demonstrate full compliance, all the time.

"As necessary", "as appropriate", "may", "normally", "should", "where applicable", "where practicable" – When any of the above terms are used, they provide flexibility of application to an organization and the degree of compliance is subject to the specific organization's interpretation.

3.8   How Often Are the ISO 9000 Standards Revised?

The original series of standards, ISO 9000, 9001, 9002, 9003, and 9004, were first issued in 1987. Subsequent revisions are always listed with the standard (e.g., ISO 9001:1994). It was the intent of ISO to review the standards every five years. The last revision cycle started in 1992 and the revisions were released in 1994. When revisions are at the final stages, they are available as Draft International Standards or DIS.

3.9   Who Has Adopted ISO 9000 Standards?

By the end of 1995, the ISO 9000 series of International Standards will have been adopted by more than a hundred countries. Table 2.2 shows the ISO 9000 standards nomenclature for major industrialized countries.

Table 2.2:   World-Wide Equivalents of ISO 9000 Standards

Country	ISO 9000 Standard Nomenclature
Australia	AS 3900
Belgium	NBN-EN 29000
Brazil	NB 9000
Canada	Z 299
Chile	NCH-ISO9000
China	GB/T 10300
Denmark	DS/ISO 9000
Finland	SFS-ISO9000
France	NF EN 29000
Germany	DIN ISO 9000
Greece	ELOT EN29000
Iceland	IST ISO 9000
India	IS 14000
Ireland	I.S./ISO 9000
Italy	UNI/EN 29000
Japan	JIS Z 9900
Mexico	NOM-CC 2
Netherlands	NEN ISO 9000
Norway	NS ISO 9000
Portugal	EM 29000
Singapore	SS 306
South Africa	SABS 0157
Spain	UNE 66-9000
Sweden	SS-ISO 9000
Switzerland	SN-EN 29000
United Kingdom	BS 5750
United States	Q9000

Typically countries that adopt the standard assign a standards name and number consistent with their existing standards. For example, the United States, which first assigned a Q90 series of numbers through the American National Standards Institute (ANSI) and the American Society for Quality Control (ASQC), in the 1994 revisions use the more complimentary numbering as the Q9000 series of numbers. The European Union (EU) has adopted the ISO 9000 series as European Norm (EN) 29000.

3.10 Where Can I Obtain the ISO 9000 Standards?

Standards are available from ISO in Geneva, Switzerland. However, in the United States it is more convenient to order from:

The American National
Standards Institute (ANSI)
11 West 42nd St.
New York, NY 10036

Tel: 212 642-4900
FAX: 212 398-0023

The American Society for
Quality Control (ASQC)
611 East Wisconsin Ave.
Milwaukee, WI 53201

Tel: 800 248-1946
FAX: 414 272-1734

4   Registrars

4.1   What Is a Registrar?

A registrar, also referred to as a "third party audit group," is an organization that is in the business of evaluating quality systems for compliance to an ISO 9000 International Standard, such as ISO 9001, ISO 9002, or ISO 9003.

4.2   What Does a Registrar Do?

The registrar conducts third party audits to assure that companies have quality systems that meet the applicable ISO 9000 series quality system standard (i.e., ISO-9001, ISO-9002, or ISO-9003).

4.3   How is a Registrar Evaluated and Accredited?

Quality system registrars are evaluated and accredited in most
countries by an accreditation body established by the national
authorities. The standard used, EN 45012, defines the general
criteria for certification bodies operating quality system
certification.   Upon approval, the registrar is accredited in the
country of the accreditation body. Registrars can be accredited by
more than one authoritative body.   Figure 2.3 shows the
relationships of accreditation bodies to registrars.

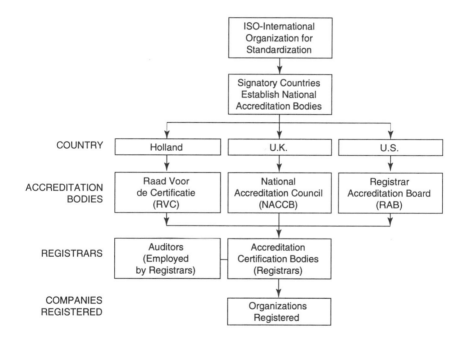

Figure 2.3   Relationships of Accreditation Bodies to Registrars

In the United Kingdom, the government is the recognition body.
The accreditation body is NACCB and it accredits registrars. The
registrar, in turn, perform audits of organization's quality systems.

Further guidance on registrars is contained in the ISO/IEC Guide 40, General Requirements for the Acceptance of Certification Bodies, and ISO/IEC Guide 48, Guidelines for Third Party Assessment and Registration of a Suppliers Quality System.

4.4 Who Are Some of the Registrars for ISO 9000?

Table 2.3 lists the names of popular quality system registrars in the United States accredited by the Registrar Accreditation Board.

Table 2.3: List of Registrars

Name and Address	Tel and Fax
ABS Quality Evaluations, Inc. 16855 Northchase Drive Houston, TX 77060-6008	Tel: 713 873-9400 Fax: 713 874-9564
Bureau Veritas Quality International Inc. 509 North Main Street Jamestown, NY 14701	Tel: 716 484-9002 Fax: 716 484-9003
Det Norske Veritas Certification, Inc. 16340 Park Ten Place, Suite 100 Houston, TX 77084	Tel: 713 579-9003 Fax: 713 579-1360
Intertek Services Corporation 9900 Main Street, Suite 500 Fairfax, VA 22031-3969	Tel: 703 476-9000 Fax: 703 273-4124
KEMA Registered Quality, Inc. 4379 County Line Road Chalfont, PA 18914	Tel: 215 822-4258 Fax: 215 822-4285
KPMG Quality Registrar 150 John F. Kennedy Parkway Short Hills, NJ 07078	Tel: 800 716-5595 Fax: 201 912-6050
Loydd's Register Quality Assurance Ltd. 33-41 Newark Street Hoboken, NJ 07030	Tel: 201 963-1111 Fax: 201 963-3299
Underwriters Laboratories, Inc. 1285 Walt Whitman Road Melville, NY 11747	Tel: 516 271-6200 Fax: 516 271-6242

In other parts of the world contact ISO headquarters or the local government authority. These are listed in the ISO catalog, or contact the commercial attaché of the country you are interested in.

Note: See Chapters 4 and 6 for further information on registrars and their selection.

5   Registration Process

5.1   What Is ISO 9000 Registration?

When an assessment of your company's quality system is conducted by a registrar and is found to be in compliance with ISO Standards (9001, 9002, or 9003), the registrar recommends registration to the accreditation agency.   The accreditation agency reviews the registrar's findings and, if satisfied, permits the registrar to issue a certificate of registration.

It is important to remember that the quality system is registered, not the product produced.

Note: The term registration is commonly used in the United States while the term certification is preferred in the European Union. The term registration is used in this book.

5.2   What Are the Steps in the Registration Process?

Typical steps in the registration process are shown in Figure 2.4:

Application

Select the ISO standard and registrar to be used.   Complete an application, providing basic information [i.e., company name, location, applicable standard, standard industry code (SIC), and statement of scope of registration].   Also include the number of employees and facility square footage affected by the scope of registration.

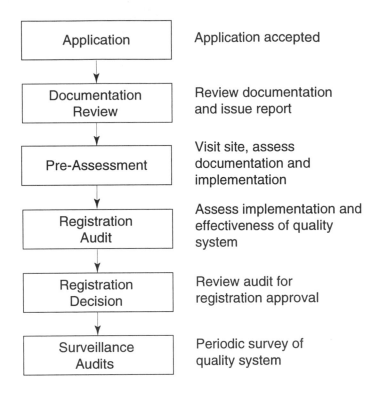

Figure 2.4   Registration Process

Documentation Review/Desk Audit

Once documentation is satisfactorily completed it is submitted to the registrar. The registrar reviews the documentation for compliance to ISO requirements. A written formal documentation review report is provided by the registrar.

Note: Some registrars prefer to perform this review at your facility. Others perform the review at their offices, saving travel costs and expenses.

Pre-Assessment (Mock Audit)

Pre-assessment or mock audit is a complete audit, determining if the quality system is ready for registrar examination. This audit can be performed by the internal audit team, a consulting team, or the registrar.

Registration Audit

Registration audit is performed when an organization believes its documented quality system is "up and running." Full registration assessment is performed on site by the registrar's auditors. Based on the results of this audit the registrar makes one of three recommendations:

- **Recommend Registration**

  This condition applies when no "major non-conformances" are found during the audit.

- **Withhold Registration Pending Corrective Action**

  This condition applies when the applicant has one or more "major non-conformances" which, in the judgment of the audit team, can be corrected by the applicant and be adequately verified by the audit team without a full re-audit.

- **Withhold Registration**

  This condition occurs when the applicant has one or more major non-conformances which, in the judgment of the audit team, require a complete audit after corrective action.

Registration Decision

The registration board typically meets monthly and evaluates all audits and their respective recommendations. Upon satisfactory completion, registration is granted and a certificate of registration is issued.

Surveillance Audit

Surveillance audits are performed by the registrar to check for continued compliance and are usually done on a semi-annual basis.

Note: See Chapter 5 for details on the various audits.

5.3   How Do I Register Multiple Sites?

It is possible to register multiple sites, but this strategy requires a lot of expertise, planning, and coordination.   For multiple site registration the registrar usually performs partial audits at each of the sites included in the scope.

5.4   What Are the Causes of Failed Registration Efforts?

Registration efforts fail for any number of reasons.   The top four reasons are:

- Lack of management support

- Lack of training

- Poor quality system documentation

- Inadequate corrective actions

5.5   How Long Does It Take to Get Registered?

The state of the existing quality system is the primary factor in determining the time required for registration.   Other factors, such as the company's commitment and the resources it is willing to expend, also play an important role.

A well-documented system may require minor changes and the registration effort may take as little as six months.   But, on the average, registration time is slightly longer than a year.

5.6 How Much Will a Registration Effort Cost?

Costs vary widely, although the internal costs will far exceed the costs of the third party activity. To those who challenge the internal costs, it is a popular view that these activities are quality related and something that the organization should have been doing all along.

External costs vary; they depend upon the size and type of facility, and normally range from $10,000 to $30,000 per facility.

6 Guidance Standards

6.1 What Are Guidance Standards?

TC/176 has developed an array of guidance standards to supplement the compliance standards.

All standards (except 9001, 9002, 9003) are called guidance standards. Some aid in the selection of the appropriate compliance standard. Others provide the proper guidance for words and terms used in the quality manual, procedures, and work instructions. They expand the meaning and direction of the systems standards with in-depth information.

Core standards specify **what** must be done. Guidance standards provide insight on **how** to do it. Core standards are also very generic and not easily understandable. Several of the guidance standards have been developed to help with this issue.

6.2 What Guidance Standards Are Available?

ISO 8402   Quality Management and Quality Assurance -Vocabulary

Many ordinary words, in everyday use, are used in the quality field in a specific or restricted manner. The intent of this standard is to clarify and standardize the quality terms as they apply to the field of quality management.

Quality Management and Quality Assurance Standards

ISO 9000-1 Guidelines for Selection and Use

This standard provides guidance for the selection and use of the ISO 9000 family of International Standards on quality management and quality assurance.

ISO 9000-2 Generic Guidelines for the Application of ISO 9001, ISO 9002, and ISO 9003

The purpose of this standard is to enable users to have improved consistency, precision, clarity, and understanding when applying the requirements of quality system standards ISO 9001, 9002, and 9003.

ISO 9000-3 Guidelines for the Application of ISO 9001 to the Development, Supply, and Maintenance of Software

The process of development and maintenance of software is different than for most other types of industrial products. This standard provides the necessary additional guidance for quality systems where software products are involved.

ISO 9000-4 Guide to Dependability Program Management

This standard provides guidance on dependability program management and covers the essential features of such programs. In management terms, it is concerned with what has to be done, why, when, and how.

Quality Management and Quality System Elements

ISO 9004-1 Guidelines

This standard provides guidance on quality management and quality system elements. It explains each element of the quality system in detail, but the extent to which these elements are adopted depends upon factors such as market being served and nature of product.

ISO 9004-2 Guidelines for Services

This part of ISO 9004 gives guidance for establishing and implementing a quality system within an organization. It is based upon the generic principles of internal quality management described in part 1, and provides a comprehensive overview of quality systems specifically for services. The concepts, principles, and quality systems elements described are applicable to all forms of service.

ISO 9004-3 Guidelines for Processed Materials

This standard is a basic guide to quality management for processed materials.

ISO 9004-4 Guidelines for Quality Improvement

This standard is a guideline for management to implement continuous quality improvement within an organization. It explains in detail the actions to be taken throughout the organization to increase the effectiveness and efficiency of activities and processes to provide added benefits to both the organization and its customers.

ISO 9004-5 Guidelines for Quality Plans

This standard provides guidance to assist suppliers and customers in the preparation, review, acceptance, and revision of quality plans.

ISO 9004-7   Guidelines for Configuration Management

This standard provides guidance on the use of configuration management in industry and its interface with other management systems and procedures.   It explains the management discipline required over the life cycle of a product to provide visibility and control of its function and physical characteristics.

Guidelines for Auditing Quality Systems

ISO 10011-1 Auditing

ISO 10011-1 emphasizes the importance of quality audit as a key management tool for achieving the objectives set out in an organization's policy.   It establishes the basic audit principles, criteria, and practices; and provides guidelines for establishing, planning, carrying out, and documenting audits of quality systems. It is sufficiently general in nature to permit it to be applicable or adaptable to different kinds of industries and organizations.

ISO 10011-2 Qualification Criteria for Quality Systems Auditors

ISO 10011-2 provides guidance on qualification criteria for auditors of quality systems.   It is applicable in the selection of auditors to perform quality systems audits called for in ISO 10011-1.

ISO 10011-3 Management of Audit Programs

ISO 10011-3 provides guidelines for management of quality system audit programs defined in 10011-1.

Quality Assurance Requirements for Measuring Equipment

ISO 10012-1 Meteorological Confirmation System for Measuring Equipment

This standard contains quality assurance requirements for a supplier to ensure that measurements are made with the intended accuracy. It also contains an explanation for the implementation of these requirements.

Guidelines for Developing Quality Manuals

ISO 10013   Guidelines for Developing Quality Manuals

This standard explains the process for the development, preparation, and control of quality manuals. It presents the documentation hierarchy and formats typically used when documenting quality systems.

Note: In addition to the above quality system standards, ISO has published a position paper entitled Vision 2000, which explains the strategy for international standards' implementation in the quality arena during the 1990s.

6.3   How Do I Use The Guidance Standards?

The guidance standards are extremely useful documents and enable the user to develop a better quality system. The conformance standards state what is required, and the guidance standards assist you with the how's.

The recommended sequence for using the standards is as follows:

1   Use guidance standard 9000-1 to select the appropriate core standard.

2   Study applicable core standard 9001, 9002, 9003 and ISO 8402.

3   Read guidance standard (9004-1) to provide quality management guidelines.

4   Read ISO 10013 to develop quality system documentation.

5   Read ISO 10011 to develop internal audit program.

6   Read other ISO standards as necessary.

7   Conclusion

This chapter should have provided sufficient information about the "World of ISO" and enabled you to understand the 5 W's and 1 H involved in planning your company's registration effort.

Chapter 3
What You Have to Have

Companies are in business to provide customer satisfaction and be profitable. These are the eventual goals of any system. It is important to recognize that this system (man, machine, material, and method (4 M's)) must operate in a proper environment (E) to achieve the goals. Consider Figure 3.1.

Top management is represented by the guitarist. The quality system is depicted by the guitar consisting of the 4 M's and E as denoted by the strings. The product is music, which is realized by the skill of management to arrange and organize the activities of the quality system to achieve a desired or effective combination of quality, cost, and delivery (QCD).

Management and the guitar are the inputs, music represents the output, customer reaction, of course, is system feedback, and the net result is customer satisfaction and profitability.

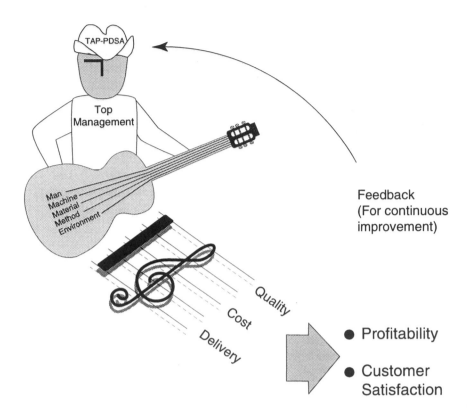

Figure 3.1   Guitarist Representation of a Company

ISO 9000 registration is an ideal way to orchestrate such a system. However, to do so effectively and efficiently, the proper framework must be established.

This chapter describes the framework required by addressing the following sections:

- Reason, Benefits, and Disadvantages of ISO 9000 Registration

- Management Commitment

- Company-Wide Involvement

- Working Environment

1 Reasons, Benefits, Disadvantages of ISO 9000 Registration

1.1 Reasons for ISO 9000 Registration

Every company has its reason for pursuing registration. The most common reasons are:

- Continuous Improvement
- Marketing Strategy
- Customer Requirement

1.1.1 Continuous Improvement

Most companies pursue registration as a part of their continuous improvement effort. Registration requires the development of a quality system which becomes an excellent starting point for continuous improvement. Continuous improvement involves company-wide activities which aim at improving customer satisfaction.

1.1.2 Marketing Strategy

Obtaining registration can be a part of a company's marketing strategy and provides an excellent sales pitch. The International Standard is recognized worldwide and thus registration provides international acceptance. Global markets open and become more accessible, providing export opportunities. This results in opportunities for higher sales. Registration also provides an edge over competition if they are not registered, and a leveling of the field if they are.

1.1.3 Customer Requirement

Many customers are requiring ISO 9000 registration of their supplier's quality system as a condition for conducting business. In such a scenario, ISO 9000 registration becomes a requirement and the supplier is left with no other option but to pursue registration.

1.2 Benefits of Registration

Besides fulfilling the specific reasons, there are other benefits acquired as a result of registration.

- Improved System Documentation
- Fewer Customer Audits
- Improved Customer Relationships
- Improved Quality Awareness

Note: However, it is important to remember that all benefits of registration will not be obtained overnight. Some of them will be realized immediately and others will accrue slowly and become fully evident only after a period of time.

1.2.1 Improved System Documentation

The ISO 9000 registration effort provides an excellent opportunity to eliminate excess documentation. This effort results in a lean and proactive quality system that contains better documents and is usually more efficient and cost effective.

1.2.2 Fewer Customer Audits

Another registration benefit is usually "an improved customer–supplier relationship." This results in fewer customer audits and provides a chance to concentrate on improving the quality system rather than chaperoning the customers around. In a particular company there were fifteen customer audits in 1992 (before registration) and only one customer audit in 1993 (after registration), thus saving time and money, not to mention reduction in paperwork (i.e., the preparation of audit reports and questionnaires).

1.2.3 Improved Customer Relationships

A side effect of registration is a strengthened relationship between customer and supplier. The customer has confidence that the supplier is quality conscious and interested in supporting the customer through continuous improvement. This, in turn, may result in fewer customer requests (i.e., QC data, tests, etc.).

1.2.4 Improved Quality Awareness

The company that practices quality soon becomes aware of the fact that good quality pays a variety of dividends. For example, marketing appreciates that manufacturing can meet its commitment to quality, price, and delivery and seeks opportunities to utilize these new found strengths. Customers return to suppliers of products that they are satisfied with. Engineering increases reliability of design, knowing that operations will consistently execute these higher requirements. Top quality companies attract the best people available. Success builds on success!

1.3 Disadvantages of Registration

There are no meaningful disadvantages of registering the quality system to ISO 9000. As a matter of fact, it is wise to develop an ISO compliant system even if the company does not wish to pursue formal registration. As one CEO put it: "It is stuff you should be doing in the first place anyway."

2   Management Commitment

Most projects that fail, do so due to lack of management commitment and direction. This is very true for the ISO 9000 registration project.

The technical committee that prepared the International Standard recognized the need for ongoing management support and participation by making it a quality system requirement (4.1 Management Responsibility)

ISO 9004-1 states: "The responsibility for, and the commitment to a quality policy belongs to the highest level of management. Quality management encompasses all activities of the overall management function that determine the quality policy, objectives and responsibilities, and implement them by means such as quality planning, quality control, quality assurance, and quality improvement within the quality system."

Management must understand the scope of the registration effort and that it requires a fair amount of financial and human resources. Top management evaluates the advantages and disadvantages, and then makes the decision to proceed with ISO 9000 registration.

Note: Our management demonstrated its support through an "all hands" kick-off meeting. All employees were gathered and the CEO presented the reasons for the company's decision to pursue registration and the impact it would have on the business strategy.

3   Company-Wide Involvement

ISO 9000 registration is not a one-man show. Every employee is affected by the registration effort and is involved in some way or other. It requires active participation by everybody within the organization, working towards a common goal, that of registration. Only visible management support can make this happen. The complete company becomes one team and works towards this common goal.

3.1 Top Management

Top management defines its policy for quality, objectives for quality, and its commitment to quality.

3.2 Middle Management

Middle management ensures that the policies and objectives of upper management are communicated and implemented down into the organization through system procedures.

3.3 Technical Staff

The technical staff develops and documents quality products and their manufacturing processes.

3.4 Supervisors

Supervisors develop area-specific instructions and ensure that these instructions are understood and implemented.

3.5 Front Line Employees

Front line workers make up the majority of the workforce and it is they who manage and operate the processes by following quality documentation and work instructions.

4 Working Environment

Imagine people walking around with gloomy faces, unhappy with their work, blaming each other for mistakes, fighting problems day in and day out, and focusing on quantity not quality. Productivity will slip considerably and morale is bound to decline. This should not occur. It is important that management remain diligent about establishing and maintaining a good working environment. People should have pride in their work and should feel that they are contributing to the "wellness" of the company.

It is obvious that a proper environment is crucial to the success of a project. ISO 9000 registration is no exception. For the registration project to be successful in the long term, it will require that all work together in harmony to achieve registration. The creation of the environment is best explained by Dr. Deming in his 14 points of management.

Deming states: "The fourteen points all have one aim, to make it possible for people to work with joy."

These are simple statements requiring deeper understanding. Their implementation requires a cultural change from short-term profitability to long-term "constancy of purpose for improvement of product and service" in most companies.

To help understand them, Dr. Deming explains each point in his own words (Courtesy: Wootten Productions) and Pulitzer Prize-winning cartoonist Pat Oliphant illustrates the points through his sketches.

4.1 Deming's Fourteen Points: A Philosophy of Life

Point #1: Create constancy of purpose for the improvement of product and service.

Pat Oliphant Illustration (Permission: Wootten Productions)

Figure 3.2   Constancy of Purpose

Point # 2:   Learn the new philosophy. Teach it to employees, to
customers, to suppliers. Put it into practice, in other
words – the new philosophy – which is one of cooperation, win-win, everybody wins.

Pat Oliphant Illustration (Permission: Wootten Productions)

Figure 3.3   Everybody Wins

Point # 3:   Cease dependence on mass inspection. Much better to improve the process in the first place so that we don't produce so many defective items – or none at all.

Pat Oliphant Illustration (Permission: Wootten Productions)

Figure 3.4   Design Quality In

Point # 4:   End the practice of awarding business on the basis of price tag alone. Instead, minimize total cost in the long run. That means one has to predict the cost of use on any product or service.

Figure 3.5   Don't Buy on Price Tag Alone

Point # 5:   Improve constantly every process for planning, production, service, whatever the activity is.

Figure 3.6   Continuous Improvement

Point # 6:   Institute training for skills. People learn in different ways, and training must take account of those differences.

Pat Oliphant Illustration (Permission: Wootten Productions)

Figure 3.7   Training for Skills

Point # 7: Adopt and institute principles for the management of people. I'm referring to the management of people for recognition of different abilities, capabilities, aspirations.

Pat Oliphant Illustration (Permission: Wootten Productions)

Figure 3.8   Institute Leadership

Point # 8:   Drive out fear, build trust. It's purely a matter of management.

Pat Oliphant Illustration (Permission: Wootten Productions)

Figure 3.9   Drive Out Fear

Point # 9: Break down barriers between staff areas. In other words, build a system. Build a system within your organization for win-win, where everybody wins. This means cooperation. It means abolishment of competition.

Pat Oliphant Illustration (Permission: Wootten Productions)

Figure 3.10 Break Down Barriers

Point # 10: Eliminate slogans, exhortations, targets asking for zero defects, new levels of productivity. Nonsense! If you don't need a method, why weren't you doing it last year? Only one possible answer: You were goofing off.

Pat Oliphant Illustration (Permission: Wootten Productions)

Figure 3.11   Eliminate Slogans

Point # 11: Eliminate numerical goals, numerical quotas for any-
body. A numerical goal or quota accomplishes noth-
ing.

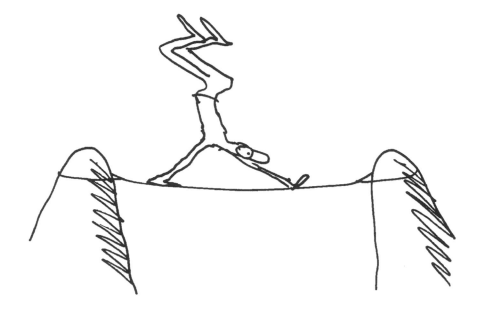

Pat Oliphant Illustration (Permission: Wootten Productions)

Figure 3.12   Method

Point # 12: Remove barriers that rob people of joy in their work. This will mean abolishing the annual rating or merit system which ranks people, creates competition and conflict.

Pat Oliphant Illustration (Permission: Wootten Productions)

Figure 3.13   Joy In Work

Point # 13:   Institute a vigorous program of education and self-improvement.

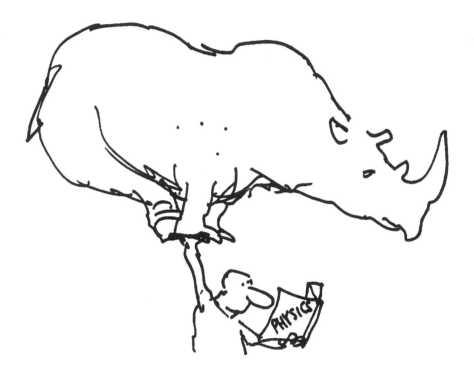

Figure 3.14   Continuing Education

Point # 14: Accomplish the transformation; that is, continue to study the new philosophy. Develop a critical mass in your organization that will bring about the transformation.

Pat Oliphant Illustration (Permission: Wootten Productions)

Figure 3.15   Accomplish the Transformation

5   Conclusion

Chapter 3, "What You Have to Have" represents sound business thinking! It explains the three elements: a reason for ISO 9000 registration, top management commitment and company-wide involvement, and a conducive working environment essential for a successful long-term registration.

These elements represent the "Get Ready" stage for your registration effort and the next chapter, "Jump Start," represents the "Go" stage!

Chapter 4
Jump Start

This chapter provides the reader the necessary first steps to get a registration effort "jump started."

Once top management has formally determined the reason for pursuing registration and expressed its commitment to the registration effort, it is time to proceed. It is important that the project be properly launched as organizations waste a significant amount of time and effort trying to find the right direction. "Jump start" prevents this situation from happening and results in a shorter overall completion time at a lower cost. "Jump start" consists of the following important activities:

- Selection of a Management Representative

- Formation of a Project Team

- Needs Assessment Audit (NAA)

- Preliminary ISO 9000 Registration Plan (PIP)

- Registrar Selection

- Measurements System

- Quality System Documentation (QSD)

Consistent with this book, the TAP approach is demonstrated here once again. Selection of the management representative, formation of the project team, and their training represents the "T", the needs assessment audit represents the "A", and preparation of the preliminary ISO registration plan represents the "P".

It is important to point out that this chapter provides only the basic information regarding jump start activities. Detail is provided in later chapters: Chapter 5, *TAP,* covers the management representative, project team, and the plan. Chapter 6, *Registrar Selection,* presents the registrar selection process. Chapter 7, *Measurements,* covers measurements and costs. Chapter 8, *Quality System Documentation,* explains the design and documentation of the quality system.

1   Selection of a Management Representative

After the decision to seek registration is made, the first important activity is the selection of a management representative by executive management.

1.1 What Is a Management Representative?

Using the words of the standard: Management Representative is a member of management who is appointed by the supplier's executive management to:

- Assure that the quality system requirements are established, implemented, and maintained in accordance with the International Standard.

- Report on the performance of the quality system to the supplier's management for review and as a basis for improvement of the quality system.

Note: The responsibility of a management representative may also include liaison with external bodies on matters relating to the supplier's quality system.

The management representative is usually the individual who leads the ISO 9000 registration effort and heads the project team.

For larger companies, the decision to have a "titular" management representative, who is a member of the senior management team, should be considered. This person will be assisted by the project manager who becomes the "de facto" or "working" management representative leading the project team and managing the day-to-day activities.

Each company should determine what is best for its particular management style. It should be recognized that the successful management representative takes on an assignment that will probably become a significant, if not total, allocation of time. If the commitment is not recognized, the registration effort will be seriously, if not fatally impacted.

1.2   What Are the Duties of a Management Representative?

It is recommended that a position description be prepared for the management representative. It also appears that the current interpretation of the International Standard requires that the "responsibilities" of the management representative be more formally described than in the past, so a position description becomes a useful document for this purpose also.

The model position description (duties, skills, knowledge, and training) for the management representative/project leader are provided in Table 4.1.

Table 4.1:   Management Representative Matrix

Duties
•   Ensure that quality system requirements are established, implemented, and maintained in accordance with the International Standard. •   Report on the performance of the quality system to the supplier's management for review and as a basis for improvement of the quality system. •   Lead ISO 9000 registration effort. Select and lead the project team. •   Be responsible for the internal audit program. •   Develop quality system documentation, implementation approach, and strategy. •   Determine and provide for company-wide ISO 9000 training. •   Registrar selection and interface process. •   Develop project cost and measurement criteria. •   Maintain the registration as needed throughout the life of the registration.
Skills and Knowledge
•   Project management and leadership skills. •   Internal auditing skills. •   Documentation skills (in-depth understanding of the family of ISO 9000 International Standards). •   Training skills. •   Registrar knowledge. •   Accounting skills.
Training
•   Team leadership training. •   Lead assessor training.

1.3   Who Should Be the Management Representative?

It is ideal for the ISO 9000 project manager to be the management representative.   In most companies the quality assurance manager usually assumes the responsibility of a management representative.

Note: Try selecting a management representative from within the company, as this has significant advantages.   Recruiting an outsider and allowing time for this individual to become familiar with your system requires a considerable amount of time which may result in significant delays.   It is also advisable to select a person who is conversant with modern quality thinking and systems thinking.

2   Formation of a Project Team

Having selected the management representative, the next task is to assemble a project team for the purpose of attaining registration.

2.1   What Is an ISO 9000 Project Team?

An ISO 9000 project team is a group of professionals representing all of the functional departments (engineering, quality, documentation, manufacturing, marketing, materials, information systems) that are within the scope of the ISO registration and are charged with developing an ISO 9000 compliant quality system.

Note: Try to select people who want to be on the team.

2.2   What Are the Duties of the Project Team?

The duties of the project team are quite similar to the duties of the project leader. Each project member has some area of specialization required for the ISO 9000 project. Table 4.2 provides a matrix of duties, skills, and training requirements for the project team.

Table 4.2:   Project Team Matrix

Duties
•   Work together with project leader to achieve registration.
•   Perform internal audits.
•   Develop quality system documentation and its implementation.
•   Determine and provide for company wide ISO 9000 training.
•   Assist in registrar selection.
•   Develop and perform project measurements and costing.
•   Help maintain the registration throughout the life of the registration.
Skills and Knowledge
•   Project Management/Team Management Skills.
•   Internal Auditing Skills.
•   Documentation skills (In-depth understanding of the family of International Standards).
•   Training Skills.
•   Accounting Skills.
Training
•   ISO 9000 Implementation and Registration Training.
•   Internal Auditor Training.
•   Documentation Training.

Note: Most companies jump straight to the Needs Assessment
Audit and pay little attention to the training needs of
management representative/project leader and the project
team. Experience has shown that an initial investment in
training the project team will help speed up the registration
process significantly.

2.3 Why Do We Need an ISO 9000 Project Team?

ISO 9000 is not a one-man show. ISO 9000 is a multifaceted,
company-wide project involving all departments. To efficiently
achieve registration, representatives from each department must
work together as a team towards the common goal of registration.

Note: In most companies, the management representative and
members of the project team will devote only a fraction of
their time to this project.

3   Needs Assessment Audit (NAA)

When the management representative and the project team (trained
auditors) have been trained, it is time to perform the Analysis step
(the A in TAP), or the needs assessment audit.

Note: The NAA results are a major decision point for management.
At this time, management receives another opportunity to
reconfirm its commitment to the registration effort.

3.1 What Is a Needs Assessment Audit?

The first comprehensive internal audit per the applicable ISO 9000
standard will be a needs assessment audit. The NAA examines
existing documentation and its implementation.

This audit is primarily performed to provide a clear, in-depth picture of the deficiencies within the existent quality system so that proper planning (P in TAP) can be done to bring the quality system into compliance.

In addition, this information is also extremely vital in determining the resources required to achieve registration (i.e., the cost/budget and man-hours).

The results of the NAA must be properly communicated to all affected. This ensures that everyone is focused in the same direction. Often a formal meeting, conducted by senior management, is an effective vehicle to communicate.

Note: The NAA itself can be used not only to audit, but also to train at the same time. This approach is very useful as it reduces the training time and standardizes the standard interpretation.

3.2 What Should You Audit?

Audit the complete quality system activities that are within the scope of the applicable International Standard (9001, 9002, or 9003).

Note: You will realize that almost all departments will be affected by the scope of registration and will need to be audited.

3.3 Who Performs the Needs Assessment Audit?

The NAA is an assignment for the project team.

It is possible, and completely acceptable, to have the needs assessment audit performed by external auditors (i.e., consultants or contract personnel), providing that the project team has evaluated the advantages and disadvantages of this approach. This approach is not discouraged, but experience shows that an initial in-house effort works best. Auditing one's own system gives a better understanding of how things really are!

Note: A conflict of interest would prohibit the registrar candidates
from participating in this activity.

3.4 How Is a Needs Assessment Audit Performed?

The trained auditor uses a systematic approach as shown in
Figure 4.1.

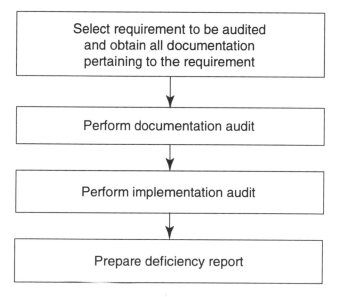

Figure 4.1   Needs Assessment Audit Flowchart

Select Requirement

Select the element to be audited and prepare a check list and obtain
the most current documentation pertaining to the element.

Perform Documentation Audit

Perform an audit of the documentation. In the audit, follow the
hierarchy of documentation. (For example, a work instruction must

implement a procedure. The procedure in turn must implement a quality policy.) The key questions are: "Have you documented what you do?" and "How does it comply to the standard?" After auditing all requirements, determine the documentation compliance level.

Perform Implementation Audit

The auditor then observes the element being performed against the standard. The key questions is: "Have you implemented what has been documented?"   After auditing all the requirements for implementation, determine the implementation compliance level.

Note: In some situations the activity being performed complies with the standard requirement but there is no supporting documentation. Consider this to be only a documentation deficiency.

Prepare Deficiency Report

Prepare a detailed deficiency report containing the overall compliance level. This report serves as an excellent baseline for measuring project progress.

4   Preliminary ISO 9000 Project Plan

After the completion of the NAA (A in TAP) the preliminary project plan (P in TAP) is developed.

4.1   What Is a Preliminary ISO 9000 Registration Plan (PIP)?

A preliminary ISO 9000 project plan identifies major milestones and their expected completion dates as shown in Figure 4.2.

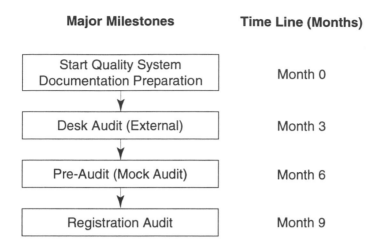

Major Milestones   **Time Line (Months)**

Start Quality System Documentation Preparation	Month 0
Desk Audit (External)	Month 3
Pre-Audit (Mock Audit)	Month 6
Registration Audit	Month 9

Figure 4.2   PIP Flowchart

Start Quality System Documentation Preparation

This step involves the development of ISO complaint documentation or in other words "Document what you do."

Desk Audit (External)

At this point, the registrar examines the documentation against the requirements of the applicable ISO 9000 standard.

Pre-Audit (Mock Audit)

This audit is usually a preparatory audit used to determine "system readiness" and correct any last-minute deficiencies. It is usually an optional activity.

Registration Audit

The registration audit is the moment of truth when the registrar performs a formal compliance audit.

Note: When developing the time line for the above flowchart keep the following key points in mind:

- It is important to block out dates with the registrar. Many companies have had to delay registration as registrars were not available to perform audits when desired.

- Significant energy must be devoted to planning. ISO 9000 is a big project and must be scheduled properly to be completed successfully.

5   Registrar Selection

The selection of a registrar should be an early priority for the management representative and the project team. The team probably has limited knowledge and experience in the selection process for a registrar. It takes time to "learn the ropes."

Note: Always remember that you are developing a long-term relationship that will probably exist as long as your company wishes to be registered to the International Standard.

5.1   Why Is Early Registrar Selection Important to the Registration Process?

- The registrar selection process becomes a learning experience since it involves dealing directly with the registrar who is thoroughly familiar with the registration process.

- Registrars are not permitted to consult, but they can often provide guidance simply by telling about their procedures and how they operate. (One of the skills that team members who work on the registrar selection process develop is how to phrase a question for information that the registrar's representative can respond to without tripping over the consulting issue.)

- The project team starts to learn about "delivery" and "costs" when the registrar discusses response time to audit requests and the costs related to these activities.

5.2 Who Will Select the Third Party Registrar?

In most companies, the registrar is selected by the management representative and project team.   Consensus is strongly recommended.

5.3 What Are the Steps in Selecting a Registrar?

Figure 4.3 outlines the major steps for selecting a registrar.

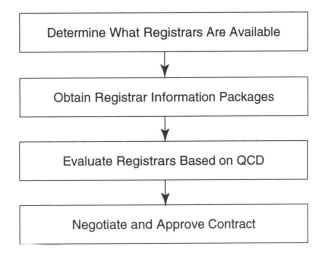

Figure 4.3   Registrar Selection

Determine What Registrars Are Available

To locate third party registrars in the U.S. contact:

Registrar Accreditation Board
c/o American Society for Quality Control
611 East Wisconsin Avenue
Milwaukee, WI 53201

Phone: 800 248-1946
Fax: 414 272-1734

In other parts of the world, contact your local government authority or the ISO headquarters in Geneva.

Obtain Registrar Information

Contact each registrar and request an information package, listing companies that the registrar has registered. These packages should be reviewed and evaluated carefully by the project team. They are the basis for the first screening out of unsuitable candidates.

Some registrars specialize in certain industries, and if they do not fit into the service or manufacture that is compatible with your company, they should be discarded. Preferred registrars should have auditors who are thoroughly familiar with your industry. The list of companies registered by the registrar will also identify if the registrar is active in your area. Learn everything about individual registrars. This involves studying how the registrar prepares a company for the preliminary audits (e.g., the desk audit, the pre-audit, and the registration audit).

Evaluate Registrars

It is strongly recommended that the sub committee for registrar selection prepare a comprehensive check list or evaluation guide based upon QCD principles. The detailed evaluation procedure is covered thoroughly in Chapter 6. The project team needs to winnow the list of third party registrar candidates down to approximately three finalists from whom the project team will select the registrar to work with.

Negotiate and Approve Contract

The project team approves the registrar candidate and initiates the negotiation process. The management representative is usually assigned this responsibility. The registrar is asked to submit a contract for approval. Upon receipt the project team evaluates the proposed contract to ensure that the all requirements have been satisfied. If the contract is satisfactory it is approved.

6   Measurement System

All project plans should have a measurement system built into them. In the words of Peter Drucker, "If you cannot measure it, then you cannot manage it." The ISO project is no exception.

6.1 What Are Measurements?

Measurements are **analyses** in TAP resulting in an improved plan and **study** in PDSA resulting in appropriate corrective action.

The ISO Measurements consist of project and system measurements at three stages, namely input, during process, and output.

6.2 ISO 9000 Project Measurements

Project measurements are measurements that are used to track the ISO 9000 project and are based on the NAA results. NAA results provide a clear picture of the systems compliance level at the start of the project against which the planned progress can be measured and tracked.

Typical ISO 9000 project measurement elements are:

*   Activities against Time as shown in the Gantt Chart (plan section in Chapter 5) (documentation compliance levels, audits, observations, and respective corrective actions, etc.)
*   Cost (money spent against budget)

6.3 Quality System Performance Measurements

Performance measurements are measurements that determine the improvements in quality system performance that resulted due to the registration.

This type of measurement consists of the "key" system quality, cost, delivery (QCD) elements indicating the health of the system. Identify these elements at the start of the project so that improvements in the quality system that resulted due to your successful registration effort can be measured.

Typical quality system performance measurement are:

- Customer returns
- Reject ratios
- Process capability indices

Note: The project and system measurements are best determined using a matrix of 4 M's and QCD as explained in Chapter 7.

7   Quality System Documentation

The preparation of quality system documentation (QSD) that is to be in compliance with the standard and its effective implementation is the heart of the registration effort. In fact, the standard states the following:

"The Supplier shall prepare documented procedures consistent with the requirements of this international standard and effectively implement the quality system and its documented procedures"

7.1   What Is Quality System Documentation?

It is documentation that describes an organization's quality system.

ISO 8402 defines "**quality system**" as: "The organizational structure, responsibilities, procedures, processes, and resources for implementing quality management." A **document** is anything that is written.

The ISO 9000 compliance standards state the following:

"The supplier shall establish, document, and maintain a quality system as a means of ensuring that product conforms to specified requirements. The supplier shall prepare a quality manual covering the requirements of the International Standard."

The ISO 9000 series of quality system standards require development and implementation of a documented quality system. Further, these standards then require the preparation of a quality manual.

7.2 How Do You Prepare Quality System Documentation?

Quality system documentation is developed in two steps (the QSD process is explained in detail in Chapter 8).

- Design the quality system
- Document the quality system

Design the Quality System

Quality system design involves mapping or flowcharting all quality system activities from design, through development, production, installation, and servicing.

Document the Quality System

Quality system documentation involves documenting the above.

Note: ISO guidance standard 10013: Guideline for Developing
        Quality Manuals, provides excellent direction on
        documentation organization and preparation.

8   Conclusion

This chapter identified and explained the first essential steps geared
towards an efficient registration effort using the TAP approach.

The information provided is drawn from real experiences and should
help launch the registration effort in the right direction.

Further information on these topics is provided in the next four
chapters.

Chapter 5
Training, Analysis, and Planning (TAP)

This chapter presents the training, analysis, and planning (TAP) requirements in detail for an ISO 9000 registration effort. The Training element describes the training requirements for all levels of an organization, the Analysis element explains the various audits at different stages of the registration project, and the Plan explains the path to registration through the development of a TAP-PDSA master plan.

1   Training

Training is the first step in a registration effort and represents both training, which can be implemented immediately, and education, which provides the intellectual insight to understand and appreciate the entire quality management system, now, and in the future. In this chapter, training and education are used in the same context.

In the TAP approach, training is multifaceted and covers the entire workforce from executive management to the first-level operations worker. The approach described is based on the idea that a successful long-term ISO 9000 registration begins and ends with training. Since registration is a continuous process, training must operate parallel to the registration. However, the training needed must be planned and defined well in advance of need.

1.1   ISO 9000 Project Training Matrix

The ISO 9000 project training matrix (Table 5.1) lists the typical training requirements of an organization to achieve ISO 9000 registration. The now familiar 5 W's approach is used.

Who   Who defines the individuals and groups who are to receive the training.

What   What describes the course work or topic.

Why   Why states the purpose and anticipated results of training.

Where   Where addresses the source of training (university/ consultant/ books, etc.).

When   When indicates the timing when training should be conducted.

Cost   Cost gives approximations in USD for budget purposes.

The matrix is generic. It is recommended that the reader compile a similar training matrix to meet specific organizational requirements.

Table 5.1: ISO 9000 Project Training Matrix

Who	What	Why	Where	When	Cost
Executive Management	ISO 9000 Executive Overview	Provide executive awareness, commitment and participation	External	First step	$500
Management Representative	Lead Auditor Training	Provides a comprehensive understanding of ISO 9000 registration program	External	After management commits to the project	$1500
Project Team	ISO 9000 Implementation Training	Provides "know-how" to implement registration project	External	During formation of project team	$500
Internal Audit Team	Internal Auditor Training	Provides skills to required for internal auditing	External	Prior to Needs Assessment Audit	$500
Document Control Representative	Documentation Training (QSD)	Provides guidance to develop ISO 9000 compliant documentation	External	Before documenting quality system	$500
Managers, Supervisors, Engineers	ISO 9000 Overview	Provide direction for documentation development and implementation	Internal	Before documenting quality system	Internal Costs
Front Line Workers	OJT and Quality Policy Training	Implement documentation	Internal	Ongoing	Internal Costs

1.2   Executive Management Training

1.2.1 Who? Top Management Policy Makers

Training for top management policy makers such as CEO, COO, vice presidents, and business unit managers.

1.2.2 What? ISO 9000 Executive Overview

The following are typical topics to be covered in an "ISO 9000 executive overview" class.

- Comprehensive overview of the background and development of the ISO 9000 standards

- Introduction and overview to the ISO 9001 International Standard

- The need and importance of executive involvement in an effective Quality Management System

- Business reasons/strategy for seeking ISO 9000 registration

- Registration economics

- Insight into the registration process

- ISO 9000 as a tool for continuous improvement

1.2.3 Why? Provides Executive Awareness

Executive management training introduces top management to the advantages, disadvantages, costs, and resources involved with the registration effort. This information is necessary for management before it buys-in or commits to the registration effort. It also represents the "voice from the outside" which is important to strengthen this commitment.

Top management provides the leadership for any registration effort. These managers need to appreciate the impact of an ISO 9000 registration effort on the quality strategy and the corporation's business plan.   The activity represents costs, benefits, and assessment of risk factors, all of which must be considered and ultimately they are responsible for the success of the registration effort. They must answer the corporation's question: "Why are we doing all this?" by identifying the purposes that drive the registration effort.

1.2.4 Where? Externally

Top management training can come from many sources (e.g., consultants, institutes, and customers).

Consultant

If top management has utilized consultants for input for upper-level decision making and business planning, training should come from this source or be recommended by this source.

Educational Institutes

Most top business schools provide seminars and short courses on the ISO 9000 International Standard and its application for senior executives to use as training.

Customers

Many large corporations, with a heavy dependence upon their supplier base, provide ISO 9000 training to supplier top management and other key employees.

Note: Some management place strong dependence on an internal staff to provide training to these senior managers.

1.2.5 When? First Step

A decision made without proper information is, at best, an uninformed decision. For this reason, top management education must start well in advance of any ISO 9000 registration effort.

1.2.6 Cost? $500

A one day seminar for senior management presented by a "name" provider will cost about $500 per person. A lesser known provider may provide similar training for the half the cost. The caveat is to know the provider and reputation. Price usually is not a major consideration for top management. Training provided by a less than "top quality" provider can have repercussions that will adversely affect the future registration effort.

1.3   Management Representative Training

1.3.1 Who? Management Representative

It is recommended that the management representative take an approved ISO 9000 Lead Auditor course. For larger companies it is also desirable that the audit program manager and the project team members consider this course.

1.3.2 What? Lead Auditor Training

This comprehensive, Lead Auditor Training, 36-hour course is designed to provide a thorough understanding and knowledge of auditing in accordance with the ISO 10000 series of standards. It focuses on intensive case history studies. The material covers additional topics, such as the registrar's auditing process, documentation structures, and systems. This training involves working in teams and provides a chance to conduct a simulated third party audit. Students are also required to pass a final examination on the 5th day for full qualification.

Table 5.2 shows a typical approved Lead Auditor course outline.

Table 5.2:   Lead Auditor Course Outline

Day	Topics
Monday	• Setting the scene, systems, structures, and scope • The purpose ands the intent of quality systems • Key areas of implementation • Documentation structures and content • Defining system scope • The process of accreditation and registration • Analysis of ISO 9000
Tuesday	• Understanding the baseline: ISO 9001 in detail • Workshops and discussions for interpretations and applications in various industries • The purpose of audit and assessment
Wednesday	• The audit and assessment process, initiation to close out (ISO 10011) • Auditor tools – flow charting, matrix planning, operational analysis • The uses and abuses of checklist • Human aspects of the audit process
Thursday	• Practical experience in applying the principles • Intensive case-study work reviewing a major set of corporate documentation and preparing findings
Friday	• Reporting, closing meeting, and the examination • Team presentations of case-study findings • ISO 9000–an international perspective • The role of the auditor in society and professional ethics • A review of the RAB Quality Systems Auditor • Certification process in the U.S. • Written Examination

Source: The Victoria Group, Fairfax, VA

Note: This course is registered by the Governing Board of the IQA International Registrar of Certified Auditors.

1.3.3 Why? Provides Comprehensive Understanding of ISO 9000 Registration Program

The reason for this training becomes apparent as one goes through the course outline. For every registration effort, there is one individual or several individuals who are the real "drivers," i.e., planners and managers of the registration effort. In order to have a successful registration effort, those selected for this task need to be the most highly and comprehensively trained in the whole body of ISO 9000 and 10000 quality standards, their purpose, and application. They must be the "voice of authority." The purpose (why?) of this training has the strongest specifics.

In short, this course provides the following information which is very critical for the registration effort:

- In-depth understanding of the family of ISO International Standards related to Quality Systems and Quality Systems Management.

- Understanding of the mechanics of the registration process.

- Selection requirements for those registrars which are most beneficial to the individual company business plan.

- Provides credentials to deal effectively with the registrar, senior management, and customers.

- Ability to manage and conduct effective internal audits of the quality system components to the requirements of an ISO 9000 International Standard.

1.3.4 Where? Externally

Almost without exception, lead assessor training must be undertaken externally. These courses are generally offered by approved providers at locations globally. Since the number of courses offered continues to grow, it is usually possible to locate a time and place convenient to the individuals to be trained. Since the training format involves working in teams, it is often beneficial to undertake this training at the same time when more than one person is selected to receive lead assessor training.

1.3.5 When? After Management Commits to the Registration Project

Lead assessor training is an excellent investment in any quality program and should be undertaken as soon the decision to seek an ISO 9000 registration becomes a strong possibility.

1.3.6 Cost? $1000

Costs are very uniform ranging from $1000 to $1500 per person (USD). Non-approved courses are offered for somewhat less. For companies with a limited budget the two day internal auditor course should be adequate.

1.4   Project Team Training

1.4.1 Who? Project Teams

The project team, consisting of representatives of each department who will develop and implement the registration plan, will need to be trained.

1.4.2 What? ISO 9000 Implementation

This in-depth course provides a thorough understanding of ISO 9000 standards – both its value and limitations. It reviews the ISO 9000 series from its background, introduction, process of registration, and certification.

Table 5.3 shows a typical outline for this course.

Table 5.3:   ISO 9000 Implementation Course Outline

Key Session Topics:

• Background and essential content of the ISO 9000 series of standards
 	
• The "value" of a Quality Management System.
 	
• Development of an ISO 9000 implementation plan
   – Cost considerations
   – Benefits of implementation
   – Total quality initiatives/ISO 9000 relationships
   – Documentation strategy
   – Registrar selection
 	
• Preparing for the registration process
   – Planning and managing the certification and registration process
   – Accreditation and certification systems in the U.S. and Europe
   – Certification readiness assessment
   – Audit characteristics

Source: Excel Partnership, Inc., Sandy Hook, CT

1.4.3 Why? Provides ISO 9000 Project Know-How

This course provides information on how to implement ISO 9000 in an organization.

1.4.4 Where? Externally

It is recommended that all initial training be provided by outside providers until the first round of training is completed, unless there is a strong internal training group available. Resist the temptation to divert registration manpower to training without a thorough evaluation of advantages and disadvantages. Often the first sign of overload and schedule slippage is when training starts to get into the way of the registration effort.

1.4.5 When? During Formation of Project Teams

One of the key activities in implementing the registration process is selection of the project team members. Each member selected should have the individual's training needs carefully assessed by the management representative or the project lead. Often training can be initiated as team members are selected. All members should be trained before the team gets down to serious work. Under normal circumstances, training should be completed for this group within two months of the decision to proceed with the ISO 9000 registration effort.

1.4.6 Cost? $99 to $500

Costs range from $99 to $500 (USD) per person.

1.5   Internal Audit Team Training

1.5.1 Who? Internal Audit Team

These auditors are responsible for managing and performing internal and subcontractor audits for the organization.

In a smaller company it is possible to get by with two trained auditors, but in a larger company additional auditors are generally needed.

1.5.2 What? Internal Audit Training

Internal audit training can vary from a single day course where the basic auditing techniques, such as audit planning, execution, and reporting, are taught to meet the minimum audit requirements to a three day internal auditing class, which is more comprehensive and includes audit supervision and management, detailed audit workshops, and audit simulations. Table 5.4 shows a typical internal auditor course outline.

Table 5.4: Internal Auditor Training Course Outline

Key Topics
• Coordinate an effective Quality Management System audit against ISO 9000.
– General audit principles/objectives: identifying opportunities for change
– Training and selecting the right people for the internal audit team
– How to construct an audit program.
– Creating a system that people trust
– Planning, scheduling, and administering the audit program
– Developing checklists
– Quality manual review process (desk audit)
• Effective auditing techniques.
– Interviewing and questioning personnel
– Selecting samples for the review during he audit
– Effectively documenting the findings
• Evaluating the significance of audit findings.
• Methods for improving communication skills during the audit presentation.
• Reporting the findings and evaluations.
• Developing and implementing corrective action programs.

Source: Excel Partnership, Inc., Sandy Hook, CT

1.5.3 Why? Provides Internal Auditing Skills

This class trains personnel to become auditors with the necessary skills to evaluate quality system compliance to the requirements of the International Standard. Remember, the compliance standard requires trained personnel to perform internal quality audits.

Auditing is one of the most important quality system measurement methods. The results of the first internal audit, better known as the needs assessment audit (NAA), provide the system compliance level and determine the scope of the registration project. A poorly conducted needs assessment can send your registration effort in a completely wrong direction. Trained internal auditors will prevent such a tragedy.

Secondly, the initial mission of the internal auditors is to bring the quality system into ISO 9000 compliance through a series of internal audits and their corrective actions.

Note: The internal audits are the most important short- and long-term continuous improvement tools.

1.5.4 Where? Externally

Initially develop internal auditors using external training resources. Later on, consider an internal auditor training program. Target having at least one auditor per department. This eliminates the need for full-time auditors (also auditors cannot audit their own areas). It then becomes possible to rotate auditors. This approach provides a fresh look each time; designated auditors tend to get a bad reputation.

1.5.5 When? Prior to the NAA

Before conducting the needs assessment audit.

1.5.6 Costs? $500

A two day internal audit seminar presented by a "name" provider will run about $500 per person.

1.6   Document Control Representative Training

1.6.1 Who? Document Control Representative

Anyone with responsibilities for documenting the quality system in compliance with the ISO 9000 series.   In most companies the engineering or quality control departments assume overall responsibility for quality system documentation. Representatives of this department should be prime candidates for some ISO 9000 documentation training.

1.6.2 What? Documentation Training

Any two day course on quality system documentation is recommended. Shown below is a sample course outline.

- Introduction to ISO 10013
- Quality system documentation (hierarchy, structure, and strategy)
- Document preparation (procedures, policies, and instructions)
- Documentation pitfalls to avoid
- Converting present documentation
- Process mapping methodologies

1.6.3 Why? Provides Documentation Development Guidance

Documentation is at the heart of the quality system and it is here where most of the registration non-conformities occur. Courses of this type will provide a clear understanding of the ISO 9000 documentation requirements and help avoid problems.

1.6.4 Where? Externally

Usually the document control representative undertakes this training externally.   The person then develops and conducts internal documentation training, specific to individual department requirements.

1.6.5 When? Before Documenting Quality System

This training must occur before the start of documentation process. It may be a good idea to take this class before performing the needs assessment; as this approach will provide a chance to evaluate the documentation needs and also prepare the documentation plan at an early stage.

1.6.6 Cost? $500

The cost varies from $500 (USD) for a one day class that covers all the basics, to $900 for a two day class where a comprehensive hand-on approach is used.

1.7   Middle Management Training

1.7.1 Who? Managers, Supervisors, and Engineers

These are employees responsible for managing the ISO 9000 project implementation in their respective functional groups.

1.7.2 What? ISO 9000 Overview

Overview training is similar to an executive overview class, with more emphasis on area-specific documentation requirements, audit preparation, and less focus on the business issues.

Shown below is a typical course outline for an engineering group.

*   Introduction to ISO 9000 series of quality system standards
    (purpose, benefits, etc.)
*   The registration process (desk audit, registration audit, etc.)
*   Understanding ISO 9001 requirements in detail

1.7.3 Why? Provide Direction for Documentation

These professionals are the core group and quality system success
largely depends on their performance. They control the process and
will be responsible for documentation preparation and
implementation. Most of the compliance problems will be dropped
in their laps.

1.7.4 Where? Internally

Most training for this group of contributors is performed internally,
because of the close relationship to the actual activities that occur.

1.7.5 When?

Before the start of the quality system documentation process.

1.7.6 How? In Groups

Train specific groups together. For example, train the engineering
group together on requirements of ISO 9001 that are applicable to
them.

Use the "everyone receives training concept." It is important that
project team members start to impart their knowledge to the
functional groups as soon as possible. This process enables the
quality culture to flow down and creates an environment of common
cause. A shift in attitudes, which is literally a new paradigm, is
important. An ISO 9000 registration effort is often a difficult path

to accomplishment, and a sense of togetherness is vital towards smoothing out the bumps in the road.

1.7.7 Cost? Internal Cost

This cost is calculated per the internal training program.

1.8   Front Line Worker Training

1.8.1 Who? Front Line Workers

Front line workers are machine operators, assemblers, technicians, inspectors, packers and shippers, and other direct personnel who add value.

1.8.2 What? OJT and Quality Policy Training

Training for this group of employees is tailored to their job requirements and typically consists of the:

- **Company Policies and Regulations**

   This is usually conducted by the human resources department during new employee orientation and includes a briefing on the "Company Quality Policy" and procedures.

- **Job Skills Training**

   This is training every employee receives to do the job and usually is in the form of on-the-job-training(OJT) or internal classroom classes using documented work instructions.

- **Work Instructions**

   These explain how a specific operation is performed and include most ISO 9000 requirements, as shown in Table 5.5.

Table 5.5: **Work Instructions Relationships**

Requirement	Example
• Inspection and testing:	Process inspection requirements
• Control of quality record:	Data sheets
• Inspection, measuring, and test equipment:	Inspection equipment used
• Inspection and test status:	Pass/fail criteria using stamps, etc.
• Product identification and traceability:	Lot travelers, etc.
• Control of non-conforming product:	What to do when product is non-conforming, filling out NCR, etc.
• Document and data control:	Use only the correct revision or approved modification
• Corrective Action:	Adjust machine if product is non-conforming
• Statistical Techniques	Use control charts to determine process fluctuations

1.8.3 Why? To Implement Documentation

This training is required for workers to be able to perform their job correctly or in other words "Do what you say" as indicated by procedures and work instructions.

1.8.4 Where? Internally

On-the-job-training (OJT) or classroom in-house training.

1.8.5 When? Ongoing

This type of training is ongoing and begins as soon as documented procedures or work instructions are available.

1.8.6 Cost?

This cost is calculated per the internal training program.

1.9   Conclusion

This section provided detailed training requirements for an organization's ISO 9000 registration effort.

The training requirements outlined will not only help achieve registration efficiently, but also help design and develop a better quality system.

We cannot stress the benefits of training enough. Training ensures:

*   A good understanding of the project so that a sound registration plan with a high probability of success can be developed.
*   That all will speak the same language, that of ISO 9000.
*   The development of highly skilled employees, who are the most important resources of an organization.

Readers will soon realize that training costs represent a sound financial investment and will be repaid many-fold during the life of registration.

2   Analysis

In the ISO 9000 registration project, audit is part of the Analysis in TAP and Study in PDSA and is one of the more valuable continuous improvements tools.

ISO 8402 defines audit as "a systematic and independent examination to determine whether quality activities and related results comply with planned arrangements and whether these arrangements are implemented effectively and are suitable to achieve objectives."

2.1   Audit Classification

Audits are of several kinds:

- First party
- Second party
- Third party

Audit techniques are primarily the same, the differences come from the scope and objectives and who performs the audits.

2.1.1 First Party Audit

A first party audit is an internal audit of the quality system. It is performed by trained internal auditors and the objective is to measure quality system compliance to the applicable ISO 9000 International Standard requirements and identify areas for continuous improvement.

2.1.2 Second Party Audit

A second party audit is an audit of the vendor or sub-contractor. The audit is conducted by the supplier and the objective is to determine if the sub-contractor is able to meet established quality requirements.

2.1.3 Third Party Audit

A third party audit is a quality system registration audit. It is an audit of the company's quality system by an independent organization such as a qualified registrar who determines whether the quality system complies to the applicable ISO 9000 standard requirements or not.

An ISO 9000 registration effort involves a variety of first and third party audits as shown in Table 5.6.

What   What explains the type of audit.

Why   Why states the purpose of audits.

Who   Who identifies the individual who will conduct the audit.

Where   Where addresses the locations and scope of audits.

When   When indicates the specific timing of audits.

Cost   Cost gives approximation in USD for budget purposes.

2.2   Needs Assessment Audit (NAA)

Note: The needs assessment audit is covered in Chapter 4, "Jump Start."

2.3   Desk Audit (External and Internal)

2.3.1 What Is a Desk Audit?

A desk audit is a review of the quality manual against the applicable ISO 9000 standard requirements.

A desk audit may be of two types:

- Internal desk audit
- External desk audit

Table 5.6: ISO 9000 Project Audit Matrix

What	Why	Who	Where	When	Cost
Needs Assessment Audit	Provides clear picture of the existing quality system	Internal Audit Team/ Project Team	On-site	After developing Internal Auditors	Internal Costs
Desk Audit					
Internal	To ensure quality system documentation meets ISO 9000 requirements	Internal Audit Team	On-site	Upon completion of documentation	Internal Costs
External	To determine if the company is ready for registration	Registrar (Lead Auditor)	Registrar site	Upon completion of documentation	$1500
Mock Audit/ Pre-Registration Audit	To prepare for Registration Audit	Internal or External Auditors	On-site	Upon completion of Documentation & Implementation	Internal Costs/ $4000
Registration Audit	To obtain registration	Registrar (Audit Team)	On-site	When Mock Audit results are satisfactory	$5000
Surveillance Audit	To sustain registration	Registrar (Lead Auditor)	On-site	Every 6 to 9 months	$1500

Internal Desk Audit

This is the final examination of the quality system documentation before it is submitted to the registrar for the external desk audit. This audit looks at all levels of the quality system documentation for compliance to the standard.

External Desk Audit

The is a review of the quality manual for compliance to the applicable ISO 9000 standard.

2.3.2 Why Is It Performed?

A key objective of the internal desk audit is to determine if procedures implement policy directives and that instructions amplify procedures.

The external desk audit is performed to determine if the company is reasonably ready for a registration audit.

2.3.3 Who Performs the Desk Audit?

The internal desk audit is performed by the internal audit team.

The external desk audit is performed by the registrar. The registrar's lead auditor will usually be selected to perform the desk audit as it contains the information that will be used to prepare the registration audit plan.

Note: Often, this is the first opportunity to interface with the external auditors and learn about their audit methods.

2.3.4 Where and How Is the Desk Audit Performed?

The external desk audit is usually performed at the registrar site. The steps shown below explain the external desk audit process.

1. Company submits quality manual to the registrar.

2. Registrar reviews how well the quality manual addresses the applicable ISO 9000 standard requirements.

3. Registrar provides the company with a formal desk audit report (Figure 5.1) identifying potential deficiencies that need to be discussed and addressed by the company.

Note: If, in the lead auditor's judgment, the company has the ability to correct any shortcomings in the described quality system within one year, he/she may recommend to "continue the registration process."

4. Company closes out open issues by providing explanations or additional documentation.

Note: Closure of issues may require an on-site visit by the registrar representative.

2.3.5 When Is a Desk Audit Performed?

Desk audit is a planned event and should be performed as soon as quality system documentation is completed.

2.3.6 What Is the Cost of an External Desk Audit?

A typical desk audit requires one auditor day and will run up to about $1500.

2.4   Pre-Registration (Mock) Audit

2.4.1 What Is a Pre-Registration Audit?

A pre-registration audit, better known as a "mock audit," is a dress rehearsal or "dry run" for the registration audit.

Desk Audit Report	
Date:	10/25/9_
Company:	Carewell Mediproducts
Location:	Ramnager Road, Naintital
Standard:	ISO 9002
Scope:	Manufacture of Disposable Syringes
Documentation:	Quality System Manual
Lead Auditor:	Gurmeet Naroola

The following desk audit is a detailed appraisal of the quality system manual. This review was conducted to determine if any significant omissions or deviations from the quality standard exist prior to a formal assessment.

Summary:

Carewell Mediproducts has stated in their application that they address all 19 elements of the ISO 9002 standard. After reviewing the quality manual, the following elements may not be applicable:

4.7   –Purchaser Supplied Product – not addressed in manual

4.19   –Service – not addressed in manual

Several other elements of ISO 9002 appear to have significant weaknesses as explained in the following review:

Figure 5.1   Desk Audit Sample Report (Page 1 of 2)

Detailed Review:

4.1   Management Responsibility

4.1.1   Quality Policy

Carewell Mediproducts' mission statement is signed by the Plant Manager:

"We will strive to deliver error-free, competitive products and services on time to meet and exceed our customer expectations."

ISSUES: None

4.1.2   Management Representative

Various organizations (marketing, purchasing, operations, etc.) and their responsibilities are described.

ISSUES:

Who is the appointed management representative?
What are his responsibilities?

4.1.3   Management Review

Formal review of the entire quality system is not addressed in the manual.

ISSUES:

What reviews are made of the system?
Are all elements reviewed?
What are the inputs?
What are the outcomes?
What records are maintained?

Figure 5.1   Desk Audit Sample Report (Page 2 of 2)

2.4.2 Why Is It Performed?

It seeks to measure completely, and in detail, whether or not the quality system that has been implemented is in conformance to the applicable ISO 9000 International Standard and is ready for a third party registration audit. The registration audit is a costly endeavor for which management needs to establish a level of confidence that the registration audit will result in a recommendation for registration.

2.4.3 Who Performs the Pre-registration Audit?

This audit can be conducted by either the internal audit team or the registrar audit team. Both approaches have their benefits.

- **Internal Audit Team**

The internal audit team can do an extremely comprehensive audit and provide a final opportunity to fix any problems that might have been overlooked.

- **External Auditors**

In the view of some, external auditors providing a fresh, unbiased perspective of the quality system will do a superior job in locating deficiencies or identifying marginal performance areas of the quality system. This approach helps to better prepare for the final audit, but the costs involved are significantly higher.

2.4.4 When Is the Pre-registration Audit Performed?

This audit is conducted once the quality system has been properly documented and implemented.

2.4.5 Where Is the Pre-registration Audit Performed?

In all areas that have been described in the scope of registration.

2.4.6 How Is the Pre-registration Audit Performed?

Perform the mock audit in the same fashion as the registrar would conduct the registration audit. Talk to the registrar and request a sample registration plan. This approach will help the company better prepare for the registration audit.

Note: If the pre-registration audit shows significant problems or non-conformances, it may be a good idea to re-evaluate final registration dates with the registrar.

2.5   Registration Audit

2.5.1 What Is a Registration Audit?

The registration audit is a third party audit which verifies that the company's quality system is documented per the International Standard requirements and assesses implementation of the procedures necessary to meet the quality standard.

2.5.2 Why Is It Performed?

The registration audit is a necessary condition to achieve registration.

2.5.3 Who Performs the Registration Audit?

This audit is conducted by the third party registrar's audit team.

2.5.4 Where Is the Registration Audit Performed?

At all locations, sites, and work areas covered by the scope of the quality system to be registered.

2.5.5 When Is the Registration Audit Performed?

When the pre-registration audit results are satisfactory and the project team has confidence in the quality system.

Note: The bottom line is that the timing of a registration audit requires careful planning so that it occurs at a time which is most suitable to both the registrar and the facility being audited.

2.5.6 How Is the Registration Audit Performed?

The actual audit comes after a great deal of preparation. The registrar's lead auditor and the company's management representative communicate with each other to establish the schedule for the registration audit. Figure 5.2 shows a typical two day/two auditor audit schedule.

From the auditee's perspective the project team should plan their participation in the registration audit in great detail. For example: have the key personal identified and prepared for each element. Also, prepare all who come in contact with the auditor by making them familiar with typical questions that may be asked and the method of response. The better the preparation, the more efficient the audit.

Opening Meeting

The audit team arrives on site. Introductions are made. The lead auditor chairs the opening meeting and states the registrar responsibilities to the company, confirms the scope of audit, presents the agenda, and explains in detail how the audit will be conducted.

The management representative identifies the escorts who will accompany the individual auditors.

Registration Audit

After the opening meeting, the audit team proceeds with the audit. Requirements such as management responsibility and quality system are audited first. The reason for auditing these elements first is that a major deficiency in either one may be sufficient grounds for terminating the audit.

Carewell Mediproducts
ISO 9002 Registration Assessment Schedule
June 22nd through 23rd, 199_
June 22nd, 199_

Times:	Activity or Location:	Element #:		Assessor Number:	
				1	2
8:00 am	Conf. Room	Opening Meeting		X	X
8:30 am	Conf. Room	Management Responsibility	4.1	X	X
10:00 am	Conf. Room	Overview of the Quality System	4.2	X	X
10:30 am	Marketing	Contract Review	4.3	X	X
11:00 am	Purchasing	Purchasing	4.5	X	
	Purchasing	Purchaser Supplied Product	4.6	X	
	Training	Training	4.17	X	
12:00 pm		Break			
1:00 pm	Documentation	Document and Data Control	4.4	X	X
2.00 pm	Quality	Internal Quality Audits	4.16	X	X
3.00 pm	Various	Corrective and Preventive Action	4.13	X	X
4.00 pm	Conf. Room	Team Caucus		X	X

Figure 5.2   Registration Audit Schedule (Page 1 of 2)

June 23rd, 199_

Times:	Activity or Location:	Element #:		Assessor Number: 1	2
8:15 am	Receiving & Incoming Inspection	Product ID and Traceability	4.7	X	X
		Process Control	4.8	X	X
		Inspection and Testing	4.9	X	X
		Inspection, Meas., and Test Equip.	4.10	X	X
		Inspection and Test Status	4.11	X	X
	Δ	Control of Non conforming Product	4.12	X	X
		Corrective and Preventive Action	4.13	X	X
	Production[a]	Handling, Storage, Pkg., and Del.	4.14	X	X
	Δ	Control of Quality Records	4.15	X	X
		Statistical Techniques	4.18	X	X
12:00 pm	Break				
1:00 pm	Resume	Same elements as above		X	X
2:30 pm	Quality	Control of Quality Records	4.15	X	
	Quality	Statistical Techniques	4.18	X	
	Cal. Lab	Insp., Meas., Test Equip.	4.10		X
3:30 pm	Conf. Room	Close Open Issues		X	X
		Team Caucus			
4:30 pm	Conf. Room	Closing Meeting		X	X

[a]The team will split up to cover different production lines.

Figure 5.2   Registration Audit Schedule (Page 2 of 2)

Next the audit team breaks up, and the auditors proceed to the areas such as purchasing, training, etc. where the activity is being performed. Later, they move on to the floor, and focus on elements such as process control, product identification and traceability, inspection and testing, etc.

Discussions about activities are directed to the individuals actually performing the work rather than through escorts unless there is a language barrier. The auditor determines where to go and what to see and whom to talk to.

While auditing, the auditors take notes of both non-conformities and conformity. For those instances where there appears to be a non-conformity, an observation is written using an observation form (Figure 5.3). For a non-conformance observation to become a finding the auditor must cite the applicable clause from the standard.

All issues identified during the desk audit that were not closed are verified. If they remain open, they become observations and findings.

All areas are audited progressively until the audit is complete. It is common for auditors to revisit areas as they determine the need for additional information.

Daily Meeting

Daily, the lead auditor reviews with other audit team members the status of the audit to ensure that it is progressing in a timely and efficient manner. It is also at these meetings that the lead auditor communicates any non-conformances to the management representative. Corrective actions taken immediately generally are not reported as findings.

Closing Meeting

At the conclusion of the audit, the lead auditor conducts an exit meeting with the auditees and presents a summary of all observations, findings, and the recommendation concerning registration. Also, at this meeting the management representative acknowledges the results of the audit.

Observation Form	Company Name:	
	Location:	
	Standard:	
	Date:	

Location of Observation:	Escort:	Observation Number:

Observation:

Auditor:	Acknowledged by:

Requirement: (element# / brief description)

Major / Minor	Documentation	☐	Implementation	☐	CAR #

Figure 5.3   Observation Form

Audit Report

The lead auditor sends a registration audit report containing a summary of the audit and copies of all findings to the company's management representative, normally within ten working days of the conclusion of the audit.

Corrective Action

If a corrective action is required the auditee provides a written corrective action plan to the lead auditor within the agreed time period (normally 30 days).

2.5.7 Cost?

A typical two auditor/two day audit will run up to about $5000.

2.6   Surveillance Audit

2.6.1 What Is a Surveillance Audit?

Surveillance audit is a scaled-down version of the registration audit. Typically it is a one day audit conducted by the lead auditor.

2.6.2 Why Is It Performed?

It is a requirement, once the company has been registered, to ensure that the registered quality system remains in compliance to the applicable International Standard.

2.6.3 Who Performs It?

The surveillance audit is usually conducted by one of the auditors who participated in the initial registration audit.

2.6.4 Where Is It Performed?

Similar to the registration audit, this audit is conducted on site.

2.6.5 When Is It Performed?

Surveillance audits are usually conducted every six months. If the quality system shows improvement and no or few observations are made, the frequency of these audits can be reduced from a six month frequency to a nine month frequency.

2.6.6 How Is It Performed?

The surveillance audit is conducted in the same manner as the registration audit. Figure 5.4 shows a typical surveillance audit plan.

Time	Activity
8:00 am	Review of current status of quality manual
8:15 am	Current use of the symbols
8:30 am	Closure and verification of open non-conformances (0:15 minutes per finding)
9:00 am	Re-evaluate element 4.2 (Quality System)
10:00 am	Re-evaluate element 4.5 (Purchasing)
11:00 am	Re-evaluate element 4.17 (Training)
12:00 pm	Break
1:00 pm	Re-evaluate element 4.13 (Corrective action)
2.00 pm	Rc-evaluate element 4.18 (Internal Quality Audits)
3.00 pm	Closing Meeting

Figure 5.4   **Surveillance Audit Plan**

- **Quality Manual Review**

  A typical surveillance audit visit begins with a review of the company's quality manual, with special attention to any major changes that have occurred during the last six month period.

- **Use of Symbols**

  Next, the usage of the ISO 9000 symbols is evaluated to confirm that they are being used properly.

- **Verification of Nonconformances**

  The auditor then verifies that corrective action for all recent nonconformances has been satisfactorily implemented.

- **Audit**

  The auditor audits the quality system elements per the audit schedule. Besides the specific elements selected, the elements corrective action and internal quality audit are usually covered during each audit. The objective is to cover all quality system elements within a three year period.

- **Closing Meeting**

  The closing meeting is held after the audit has been completed and the auditor summarizes his findings.

2.6.7 How Much Does It Cost?

A one day, one auditor surveillance audit will cost about $1500.

2.7 Conclusion

This section provided the reader with the necessary information to effectively prepare for and conduct the several audits that occur during the ISO 9000 registration project. Shown are examples of actual audit plans, forms, and reports that provide as realistic a picture as possible of actual events.

In closing, the authors would like to re-emphasize that audit is part of the Analysis in TAP and Study in PDSA and is one of the more valuable continuous improvements tools.

3   Planning

Planning is the key to any successful registration project and is the centerpiece of the TAP-PDSA approach. Good planning will take the company from point A to point B in the most efficient manner. Planning should be as detailed as possible. Remember, "The devil is in the detail." Figure 5.5 represents the overall ISO 9000 registration plan using the TAP-PDSA approach.

Comprehensive information on each activity shown in the plan is provided in various parts of the book.

3.1   Training and Education

Phase one of the registration plan is training and education, and involves getting the top management, management representative, and project team trained and educated in the ISO 9000 registration effort. See Section 1 of this chapter for details regarding training.

3.2   Analysis

Phase two is the needs assessment audit (NAA), which provides a status quo/clear picture of the existing system. NAA is explained in detail in Chapter 4, "Jump Start."

3.3   Plan

Using the NAA information, the company develops the ISO 9000 registration implementation plan for the key activities (registrar selection, measurement system, QSD) as explained in Chapter 4. Further information regarding registrar selection, measurement system, and QSD is provided in Chapters 6, 7, and 8 respectively.

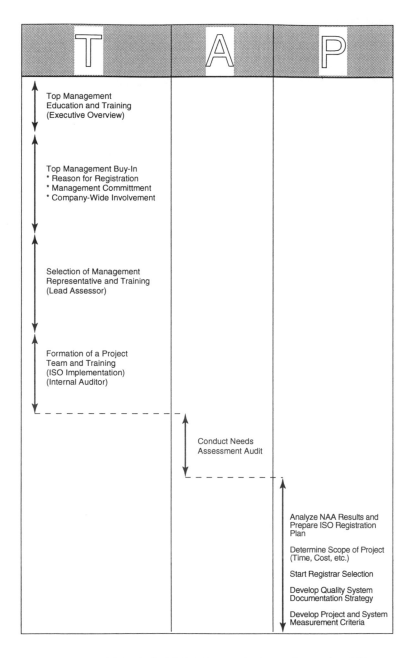

Figure 5.5   ISO 9000 TAP-PDSA Project Plan (Page 1 of 2)

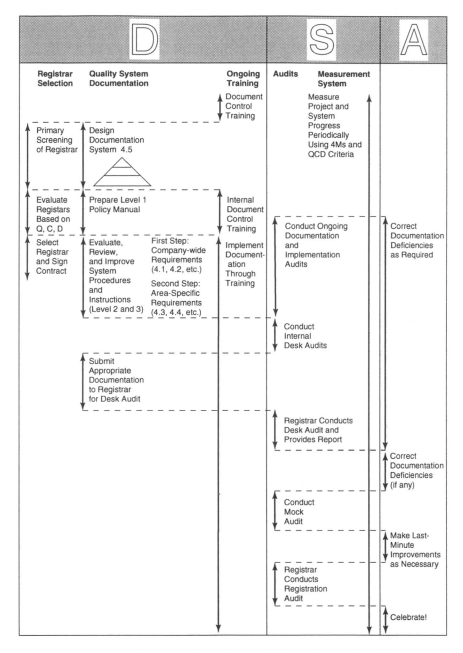

Figure 5.5   ISO 9000 TAP-PDSA Project Plan (Page 2 of 2)

3.4   Do

The plan is then implemented through the "do" step. This do step represents the implementation of the plan.

3.5   Study

The project progress is checked through ongoing audits (internal and external), and other project measurements criteria. See Section 2 in this chapter regarding details on audits and Chapter 7 for further information on measurements.

3.6   Act

Any changes to the project are conducted through ongoing corrective actions/improvements which represent the "act" stage.

The two cycles (TAP and PDSA) continuously evolve the quality system that results in an efficient ISO 9000 registration project.

3.7   Conclusion

Good planning makes the difference!

Remember, a plan is only a prediction with a probability of success, the probability of success being directly proportional to the amount and quality of training invested in the project.

Train and analyze the status quo before making a plan. Surely the project will then be successful.

Chapter 6
Registrar Selection

Chapter 4 explained the importance of early registrar selection and the selection steps. This chapter covers, in detail, the registrar evaluation process using the quality, cost, and delivery criteria.

This evaluation is usually done in two steps:

- Preliminary screening

- Final Screening

1   Preliminary Screening

Preliminary screening involves the examination of all registrars available, to narrow the list down to the top few. The task is accomplished using the following questions:

Chapter 6

Q1) Does the customer accept the registrar's accreditation?

There is no point in selecting a registrar who will not be recognized by the customer. The customer may require the use of a registrar accredited by a specific accreditation body. Some registrars are accredited by several accreditation bodies or have multiple memoranda of understanding (MOUs) and can be of great advantage if the company markets products globally.

Note: Also remember if the product is marketed in the European Union and is a regulated product under an EU directive, the notified body certifying the particular product directs which quality system registrar is acceptable.

Q2) Does the registrar perform registration to the company's standard industry code (SIC)?

The registrar should have sufficient expertise in registering companies in the SIC code of your company. Request the registrar to provide a list of companies they have registered classified by SIC code. It is important to select a registrar who is familiar with the company's type of business.

Q3) Does the registrar have acceptable credentials and references?

"By their reputation they shall be known." Call the management representatives of companies the registrar has registered. This is the most effective and expeditious method of learning about and screening registrar candidates. This important step is often ignored.

Q4) Is there a conflict of interest?

ISO/IEC guide 48 states: "An organization that advises a company on how to set up its quality system or writes its quality documentation should not provide assessment services to that company unless strict separation is achieved to ensure that there is no conflict of interest."

Companies should do the necessary research to ensure that there is no conflict of interest as this alone could jeopardize the company's registration effort.

Q5)   Is the Registrar available?

The registrar must be available for the planned registration audit dates. Demand for registrars is heavy and one should not expect a registrar to be available on short notice.

2   Final Screening

Once the top few registrars have been determined, it is time to select the registrar who best meets the company's needs.

Use the QCD criteria to select the best registrar. Remember, any fool can achieve the best measure in two out of three. Get a balance of all three to obtain the best "value" for the registration dollar.

2.1 Quality Criteria

Q1)   Does the registrar have a good quality history? (How long have they been around? How many companies have they registered? etc.)

During the supplier selection process, the first step generally is to check the quality history of the supplier in question. Treat the registrar in the same way. Ask for the quality manual, and check the non-conforming and corrective action records. Also check with the accreditation body to find out if any complains have been filed against them. Rank the registrars by both the length of time they have been in business and the number of registrations they have done in the appropriate SIC code.

Note: Many companies ask the registrar candidates to complete a supplier survey questionnaire.

Q2)   What and how many MOUs does the registrar have?

The more the better! Ensure that the MOUs cover the countries the product is marketed in.   Consider future customer requirements when assigning the rank.   Remember, this is also an important consideration for product certification.

Q3)   What is the caliber of the registrar's auditors?   Are the auditors assessors or lead assessors?

The strength of any registrar is in the caliber of the audit staff. Ask for auditor backgrounds and check their experience based on number of audits in the appropriate SIC code and number of years of audit experience. Ensure that the registrar staff consists of a high percentage of permanent auditors.

Q4)   Does the registrar treat the company like a customer? How comfortable is the company with the registrar's personnel and methods?

Is the registrar "user" friendly?   Check very   carefully for a "reading" on how registrars conduct business with their clients. A company will have a business relationship with the registrar for a number of years and it should be based upon trust, mutual respect, dependability, flexibility, and a genuine interest in serving the needs of the client.

Q5) Does the registrar understand the way the company conducts business?

The registrar must take time to understand the company's type of business especially in a fast-changing environment such as high-technology industries where products, processes, and methods are continually being changed and improved sometimes within months.

Q6) How are registrar complaints handled?

The registration process is a two-way street and should provide for a third party problem resolution process. The registrar should remember that the company is the customer and examine the problems from the company's point of view.

Q7) What happens if the registrar goes out of business?

Similar to a supplier selection, some level of confidence that the registrar will be around in future years is needed. This is done through the examination of the registrar's registration count and market share. This is one scenario where the registrar's MOUs are an advantage.

2.2 Cost Criteria

Q1) What is the registrar's total cost of registration (three years)? and does it meet the company's budget?

Cost of registrars varies considerably and involves significant expenditure of money. The two largest cost factors are labor costs (auditors and reports) and logistical costs (transportation, meals, lodging). While calculating the costs, consider the total registrar costs and not just the initial costs. There is a lot more to this than meets the eye.

Consider the following activities while determining the total registrar costs:

- Application fee

- Preparation and initial visit

- Desk audit

- Pre-audit

- Registration audit

- Certificate

- Surveillance audit

- MOUs

Note: Remember the registrar is a supplier and cost negotiation is recommended. Registrars will lower fees considerably to get business.

Q2)   Is the registrar cost structure fixed for the life of registration?

Negotiate a contract containing firm cost for a period of three years.

Note: Registrar costs should decrease as the availability of registrars increases and business becomes more competitive.

Q3)   How can a contract be canceled?

Learn the contract cancellation policies of the registrars. At some point an organization may decide to choose another registrar or give up registration for any number of business reasons. The costs associated with contract cancellations should be minimal.

2.3 Delivery Criteria

Q1) Is the registrar flexible regarding audit dates?

Registrars with extensive backlogs tend to provide limited opportunities for rescheduling. Even the best schedules sometimes require changes and the registrar should be able to accommodate without undue difficulty. In such a scenario, this criterion becomes an important factor. The ideal registrar is available on demand. Try finding a registrar who most closely approaches "availability on demand."

Q2) Does the registrar have a local office and auditors?

Having a local registrar and auditors represents significant benefits:

- Lower auditor cost (transportation/airfare/ hotel etc.).

- It is possible to establish a closer relationship with a local organization. It also gives the registrar an opportunity to learn more about the company's type of business and provide enhanced service.

- It provides an opportunity to meet and talk to the auditors and get to know their auditing attitudes and techniques.

- The local office representative can be a great networking source for providing training sources, registration information, and consultants available.

Q3) How does the registrar accommodate changes in scope or content of registration?

Changes in scope may require an amendment to the contract. Discuss the company's business direction and its effect on the scope of registration especially if new sites and new products are added on.

Some registrars may require a complete re-registration, while others may require only a change in the scope statement.

Q4) What is the registrars re-certification cycle and surveillance audit frequency?

Re-Certification

Today, most registrars do not require re-auditing of the entire system after the three year registration period. Certification holds good as long as the company continuously passes the surveillance audits. Select a registrar that does not require the complete system to be audited every three to five years. ISO 9000 registration should not become a nuisance.

Surveillance Audit

Different registrars audit with different frequencies. Some audit the entire system once a year, while others audit the system over a period of 2 or 3 years. The latter approach is preferred.

Note: Some registrars will even lower the surveillance audit frequency if the quality system demonstrates stability and continuous improvement. "The better the system, the lower the frequency of surveillance audits."

Q5) How does the registrar audit against changes to the ISO 9000 Standard?

Each accreditation organization lays down the framework and each registrar works within the framework. Individual companies negotiate compliance with their registrars. Usually a convenient surveillance date is selected as the cut-off. Some registrars may elect to dedicate a special one day audit to review the changes. The bottom line is that the registrar must give their clients sufficient time to comply to the changes and the cut-off date must be mutually agreeable.

Q6) Can a company get the same auditors every time and can they select auditors?

It is to the company's advantage to have audits performed by the same group of core auditors each time.

Less time will be spent establishing communications, understanding auditor audit methods, and explaining the complexities of the quality system.

An auditor with a previous understanding of the company's quality system will conduct the audit more skillfully. Remember, it is desirable to stabilize the registration process as much as possible.

Q7) What is registrar's response time to the desk audit and other reports?

Most registrars are good about meeting time schedules. A commitment on how fast the registrar will get audit reports back is necessary. Delayed reports, especially during the desk audit phase, can disrupt the company's registration timeline. Turnaround time for all reports should be built into the contract. Registrars usually state turnaround time in their quality manual.

Q8) How does the registrar accommodate consultants?

Many companies use consultants to help them in their registration efforts. A good question to ask the registrar is in what capacity the consultant will be allowed to participate during the registration audit. Many registrars disallow consultant presence; others will allow consultant presence, but not participation, and yet others may allow total participation. Consider registrars who accept consultants in the way the company wishes to use them.

Note: We have known companies that use consultants to perform their internal audits. This is perfectly acceptable and the registrar should not have any problems.

3   Registrar Rating Sheet

Figure 6.1 shows a Registrar Rating Sheet developed using the QCD
criteria to aid in the registrar selection process.

Criteria		Registrars			
Quality:		**A**	**B**	**C**	**D**
Q1)	Does the registrar have a good quality history? (How long have they been around? How many companies have they registered? etc.)				
Q2)	What and how many MOUs does the registrar have?				
Q3)	What is the caliber of the registrar's auditors? Are the auditors assessors or lead assessors?				
Q4)	Does the registrar treat the company like a customer? How comfortable is the company with the registrar's personnel and methods?				
Q5)	Does the registrar understand the way the company conducts business?				
Q6)	How are registrar complaints handled?				
Q7)	What happens if the registrar goes out of business?				
Cost:					
Q1)	What is the registrar's total cost of registration (three years)? and does it meet the company's budget?				
Q2)	Is the registrar cost structure fixed for the life of registration?				
Q3)	How can a contract be canceled?				
Delivery:					
Q1)	Is the registrar flexible regarding audit dates etc.?				
Q2)	Does the registrar have a local office and auditors?				
Q3)	How does the registrar accommodate changes in scope or content of registration?				
Q4)	What is the registrar's re-certification cycle/ and Surveillance Audit frequency?				
Q5)	How does the registrar audit against changes to the ISO 9000 Standard?				
Q6)	Can a company get the same auditors every time and can they select auditors?				
Q7)	What is registrar's response time to the desk audit and other reports?				
Q8)	How does the registrar accommodate consultants, etc.?				
	Section Total				

Figure 6.1   Registrar Rating Sheet

- Take the top few registrars based on the initial screening and rate them under the individual QCD criteria taking into account the explanations provided earlier in this chapter. Use any rating system (1,2,3 etc.).

- Compute individual quality, cost, and delivery totals and apply importance factor as appropriate.

- Compare and select registrar based on the best balance of QCD requirements.

4 Conclusion

Registrar selection is a crucial process in the registration effort and a significant amount of energy should be devoted to it.

This rating system provided in this chapter was used for registrar selection by the authors, in their individual ISO 9000 efforts. The list used to make up the rating system is by no means complete and it is suggested that it be modified to meet the company's specific requirements.

Good luck in the registrar selection process.

Chapter 7
Measurements

Measurements are the following:

- Analyze (A) step in TAP leading to an improved plan (P)

- Study (S) step in PDSA resulting in appropriate corrective actions (A)

During the ISO 9000 registration project, measurements are often overlooked and not given the attention they deserve. For example: When people are asked about their ISO registration project success stories, the answers are too familiar: "We improved our quality system; communication increased" and so on. These answers are too general to represent any measures of success. Success must be quantified in "hard numbers" to be meaningful. Consider the answers to be: system reject ratio decreased by 21%, documentation reduced by 31%, and customer audits reduced from 18 to zero. Now, that is measurable success.

The ISO 9001 standard addresses measurements by requiring the supplier's management to review the quality system at defined intervals, ensuring goals and objectives are met.

ISO 9004 provides further guidance on measurements by explaining that for an organization, there is a business need to attain and maintain the optimum balance of QCD (quality, cost, and delivery) utilizing the 4 M (man, machine, material, and method) resources available to the organization.

The ISO 9000 registration project must be measured and managed to keep it on track. In addition, the system must be continuously measured to ensure the ongoing success of the established quality system.

1   ISO 9000 Registration Project Measurements

The ISO 9000 registration project, like any other project, must be completed on time, within budget, and should achieve the goal of successful registration. To accomplish this goal, the project team should develop a set of project measurement criteria. These are best developed using a matrix built around the 4 M's and QCD as illustrated in Figure 7.1.

The project measurement matrix consists of twelve interrelated cells, covering the entire quality system. Each cell indicates a possible measurement that can be used to track the project progress.

Note: All components of the system are interrelated and so are the measurements. However, the ISO 9000 registration project is man and method oriented, and that is where most of the measurements criteria will appear.

	Man	Machine	Material	Method
Quality	• ISO 9000 Training • Auditing Hours • Project Planning	• Number of Uncalibrated Equipment • Number of Process Changes	• Vendor Qualification	• No. of Procedures • Documentation Compliance Level • No. of Non-Conformances
Cost	• ISO 9000 Training Costs (Internal and External) • Auditing Costs • Planning Costs • Registrar Costs • Consultant Costs	• Calibration Costs • Process Change Costs	• Vendor Qualification Costs	• System Development Costs (Management Responsibility, Quality System, . . . Statistical Methods)
Delivery	• Project Schedule (Activity against time) • Training Completion • Auditing Completion	• Calibration Compilation • Process Change Completion	• Approved Vendor List (AVL)	• System Completion (Documentation and implementation Compliance Levels)

Project Objective:
• Successful ISO 9000 Registration

Figure 7.1 ISO 9000 Project Measurement Matrix

For example: The man-cost cell contains training, auditing, and planning costs involved with the ISO 9000 registration project. For the purpose of illustration, training costs are explained. The training cost budget is estimated using the 5 W's and 1 H training matrix as explained in Chapter 5. These costs are monitored by periodic measurement of the actual plan resulting in appropriate corrective action or modification.

The matrix shown is generic.   Develop a matrix of suitable measurements that meet the needs of your organization.

2   Quality System Measurements

The end objective of the quality system is to produce an output with an acceptable combination of quality, cost, and delivery that will result in customer satisfaction and profitability.

Shown in Figure 7.2 is a quality system measurement matrix similar to the measurement matrix. The only difference is the measurement criteria and scope.

The system measurement matrix uses the same twelve cell approach as the project measurement matrix. Each cell indicates a possible measurement that can be used to track the system output or the "health" of the system, which, in turn, relates to customer satisfaction and profitability.

Some companies use these techniques to develop customized quality indices, depicting the overall health of the system.

System Objectives:
- Customer Satisfaction
- Profitability

	Man	Machine	Material	Method
Quality	• Employee Morale • Employee Turnover • Worker Output/ Productivity • Training Hours per Employee • Number of Suggestions per Employee • Number of Complaints • Absenteeism	• Machine Efficiency • Machine Capability (Reproducibility and Repeatability) • Machine Flexibility	• Reject/Scrap Rate (ppm) • Number of Non-Conferences (Internal and External) • Lots Defective • Customer Returns • Process Capability Indices (CPK) • Average Outgoing Quality Level (AQL)	• Organization Levels • Audit Non-conformances • Number of Document Changes • Number of Engineering Changes • Corrective Action Effectiveness
Cost	• Sales per Employee • Recruiting Costs • Training Costs • Direct and Indirect Costs • Unplanned Overtime Costs • Absenteeism Costs	• Downtime Costs • Repair Costs • Spare Parts Costs • Equipment Costs • Maintenance Costs	• Scrap Costs • Rework Costs • Inventory Costs • Material Costs	• Document Control Costs • System Costs (MIS, MRB) • Activity-Based Costs (ABC)
Delivery	• Number of Meetings • Complaint Resolutions • Average Employment Duration	• Repair Time • Space Availability • Maintenance Response Time	• Just-In-Time (JIT) • On-Time Shipments • Number of New Customers and Suppliers • Material Shortages • Rework Time	• Corrective Action Completion Time • Document Release Time • Information Response Time

Figure 7.2   ISO 9000 System Measurement Matrix

3   Conclusion

The value of measurements in managing the quality system cannot be overemphasized.   "Measurements are an integral part of management's effort for continuous improvement."

The measurement system should continually be evaluated, each time determining if it is providing management with the desired key information to keep the business in control.

Chapter 8
Quality System Documentation

A number of surveys have indicated, without exception, that most ISO 9000 registration efforts fail because of problems with quality system documentation (QSD). The quality system compliance standards are most specific on one point: an organization must thoroughly document what it does. Good documentation is the heart and soul of any quality system.

In the early days, there were many situations where segments of the process were performed repetitively for a long time by the same workers and the need for documentation was minimal. In the craftsmen era, the entire process was carried around in the master's head and information was doled out to the apprentices on a need-to-know basis of rote instruction.

No longer!

In today's rapidly changing global business environment, systems and processes are changing continually and it is impossible to remember anything but the most critical bits of information. Everyone involved in the quality system, from top management to frontline worker, manages with information and this information is in the form of documents. Envision a world without documents. Impossible!

The ISO 9000 series of standards are the de facto standard used in the world today because they create a format for consistency of quality. This is possible only when the quality system is fully, carefully, and correctly documented.

This chapter explains how to prepare a documented quality system consistent with the requirements of the International Standard by addressing the following questions:

• Why do we need QSD?

• What are the ISO 9000 QSD requirements?

• How is QSD organized?

• Who prepares QSD?

• How do we develop QSD?

• How do we implement QSD?

1 Why Do We Need Quality System Documentation?

Quality system documentation is needed for a variety of reasons. The most important ones are:

• **Communication**

It is a vehicle to communicate the company's quality management system.

- **Implementation**

  QSD provides specifics for implementation of quality-related activities.

- **Provides Control**

  Specifically states the criteria of acceptance.

- **Provides Consistency**

  QSD provides consistency during changing environments.

- **Trains Personnel**

  QSD assists in training individuals in how to perform their task.

- **External Purposes**

  Provides guidelines for conducting business.

2   What Are ISO 9000 Quality System Documentation Requirements?

The International Standard itself very correctly defines the quality system documentation requirements by stating:

"The supplier shall prepare documented procedures consistent with the requirements of the International Standard and the supplier's stated quality policy."

In ISO 9001 there are 20 requirements that a quality system must comply to. Their documentation requirements are:

- **Management Responsibility**

  Requires procedures that specify how management defines and implements its policy, objectives, and commitment for quality, assigns personnel to be responsible, and holds periodic reviews of the system.

- **Quality System**

  Requires the quality system to be documented to ensure product conformance to specified requirements.

  This requirement also mandates the preparation of a quality manual that covers the requirements of the applicable International Standard.

- **Contract Review**

  This requirement states that a documented procedure shall explain how a contract between the customer and supplier is reviewed and established.

- **Design Control**

  Requires procedures for: design and development planning, organization and technical interface, input, output, review, verification, and validation activities.

- **Document and Data Control**

  The document covering this requirement must explain the criteria for the document's approval, issue, and change.

- **Purchasing**

  Requires the establishment of procedures to ensure that purchased product meets specifications and is obtained from qualified suppliers.

- **Control of Customer-Supplied Product**

  Requires procedures for verification, storage, maintenance, and use of customer-supplied product.

- **Product Identification and Traceability**

  Requires procedures for identifying product throughout the system.

- **Process Control**

  Requires procedures for identifying and planning the production, installation, and servicing processes that directly affect quality.

- **Inspection and Testing**

  Requires procedures that explain the receiving, in-process, and final inspection and testing requirements.

- **Control of Inspection, Measuring, and Test Equipment**

  Requires supplier to establish documented procedures for controlling, calibrating, and maintaining inspection, measuring, and test equipment.

- **Inspection and Test Status**

  Requires procedures for identifying material status at all stages in the system.

- **Control of Nonconforming Product**

  Requires a procedure to identify, review, and disposition non-conforming material.

- **Corrective and Preventive Action**

  Requires a procedure for analyzing causes of non-conformances, and implementing appropriate corrective and preventive actions.

- **Handling, Storage, Packaging, Preservation, and Delivery**

  Requires documented procedures for handling, storing, packaging, preserving, and delivering product to prevent damage and deterioration.

- **Control of Quality Records**

  Requires the establishment of a procedure to identify, collect, index, access, file, store, maintain, and disposition all quality records pertaining to the activities of the applicable International Standard.

- **Internal Quality Audits**

  Requires procedures for planning and conducting internal quality audits to determine if the quality system is effective and complies with the planned arrangements.

- **Training**

  Requires a documented procedure for identifying training needs and providing training to all employees performing duties that affect quality.

- **Servicing**

  Requires procedures for performing, verifying, and reporting that servicing meets the specified requirements.

- **Statistical Techniques**

  Requires procedures to: identify the need, implement, and control statistical techniques required for verifying process capability and product characteristics.

This family of documents, when assembled together, form a quality systems manual.

Note: Remember that not every clause in the standard will necessarily apply to your company. Whole sections can be written as N/A but always provide convincing and clearly stated reasons for doing so.

3   How is Quality System Documentation Organized?

Most quality system documentation is organized in 4 levels as shown in Figure 8.1.

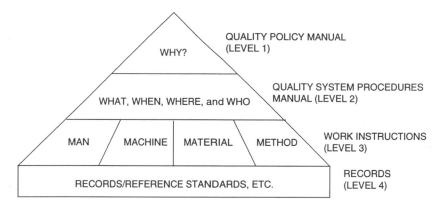

Figure 8.1   Quality System Documentation Organization

3.1 Quality Policy Manual/Level 1

This manual addresses the quality system and briefly covers all applicable policy requirements of the quality system standard selected by the organization. Here the "Why are we doing all this?" is stated. The quality policy manual is normally prepared by the quality assurance department. It is short and concise, does not contain any proprietary information, and is usually kept to one page per requirement. This is the manual that is sent to the registrar for the desk audit and is also available to new or existing customers.

3.2 Quality System Procedures Manual/Level 2

ISO 8402 (Vocabulary) defines procedure as: "A specified way to perform an activity." These procedures provide direction on how the company's quality policy is implemented. They explain in detail how each of the International Standard requirements is satisfied and it is here the what, when, where, and who questions are answered. These procedures may contain proprietary information. Preparing this documentation is usually the responsibility of individual departments. The auditors' initial focus is on this level of documentation as they make up the nuts and bolts of the quality system.

3.2.1 Structure of Level 2 Procedure

A good level 2 procedure usually contains the following:

- **Purpose and Scope**

  Defines the why, reason for, and areas impacted by the procedure.

- **Responsibility**

  States the organizational unit responsible to implement the document to achieve the stated purpose.

- **Procedure**

  Lists, step-by-step, the details of what needs to be done in a logical sequence.

- **Flowchart**

  Shows a graphical representation of the procedure.

- **References**

  Identify which supporting documents or forms are associated with this document.

- **Records**

  Identify which records are generated as a result of using this document and where and how are they retained and for how long.

3.3 Work Instructions/Level 3

Work instructions provide direction on how the procedure is to be implemented or how its requirements are to be accomplished. Work instructions define how to make, inspect, maintain, protect, transport, store, and repair products.

A good level 3 work instruction always references the man, machine, material, method, and environment (4 M's and E). It is here the technical details, forms, and data sheets are included. These instructions are usually prepared by engineers and supervisors, and should include operator input.

Note: Auditors mostly question the operators when auditing level 3 documents.

3.4   Records/Standards/Level 4

Level 4 documents consist of reference material such as specifications, standards, and records.   The ISO standards themselves can be considered level 4 documents.

Table 8.1 illustrates the various levels of documentation and their linkage. The example is for the contract review requirement.

Table 8.1:   Documentation Levels Example

Policy:	Contract Review — A flowchart procedure is in place to ensure that the requirements of the customer are fully understood, confirmed back to the customer, and mutually accepted in writing by all concerned prior to scheduling of the order. Exceptions require executive approval.
Procedure:	Procedure #323-184 — Orders against existing contracts are received from customers at New York, Houston, or St. Louis locations by telephone, fax link, or e-mail by the customer representative and are routed to the designated service representative for processing. Further processing shall follow work instructions developed for each customer.
Work Instruction:	W.I. #800-123 — Order Entry Process for Jones Account: Enter Jones Account Number on order entry terminal, follow menu displayed. Check for customer special instructions on menu file. Enter proper special instructions code into data base if instructions provided by the customer agree with file information. Order entry process must respond to all menu questions. Reserve space in production schedule and fax or e-mail request for conformation to customer. When customer acknowledges, book the order and confirm back to customer with order number.

Note: Usually quality system documentation is prepared at different levels depending on the structure and size of the company.

4 Who Prepares Quality System Documentation?

Traditionally, the quality department prepared procedures and provided them to other departments for implementation. This approach resulted in the development of an ineffective quality system as the communication was one way. Many non-conformances came from the fact that the departments had learned to develop the best procedures using their own methods.

Today, preparation of documentation should be a shared responsibility, with department personnel documenting their respective area requirements as explained by the "Document Owner and User Partnership"

4.1 Document Owner and User Partnership

4.1.1 Document Owner

The **document owner** is the department responsible for documenting the requirement. The document owner is also responsible for the implementation and upkeep of the documented requirement.

4.1.2 Document User

The document **users** are the departments affected and responsible for using the document prepared. Document owners are usually the biggest users.

The document owner and user partnership ensures the sharing of Quality System Documentation preparation by dividing the documentation requirements among the owners. Table 8.2 lists typical owners and users.

Table 8.2: Quality System Requirement Owners and Users (typical)

		TOP MANAGEMENT	QUALITY	ENGINEERING	OPERATIONS	MATERIALS	SALES	H.R. (OTHERS)
Management Responsibility	4.1	O	U	U	U	U	U	U
Quality System	4.2		O	U	U	U	U	U
Contract Review	4.3			U		U	O	
Design Control	4.4	U	O	U	U			
Document and Data Control	4.5		U	O	U	U	U	U
Purchasing	4.6		U	U	U	O		
Control of Customer-Supplied Product	4.7		U		U	O		
Product Identification and Traceability	4.8		U	U	O	U	U	
Process Control	4.9		U	U	O			
Inspection and Testing	4.10		O	U	U			
Control of Inspection and Test Equipment	4.11		U	O	U			
Inspection and Test Status	4.12		O	U	U	U	U	
Control of Non-conforming Product	4.13		O	U	U	U	U	
Corrective and Preventive Action	4.14		O	U	U	U	U	
Handling, Storage, Packaging, and Delivery	4.15		U	U	O	U		
Control of Quality Records	4.16		O	U	U	U	U	U
Internal Quality Audits	4.17		O	U	U	U	U	U
Training	4.18		U	U	U	U	U	O
Servicing	4.19				O	U	U	
Statistical Techniques	4.20		U	O	U			

Note: These are typical examples and your organization may be structured differently.

Additional benefits of this approach are:

• Company-wide involvement

• Faster project completion

• Development of an effective system

• Documentation is kept current

4.2 Document and Data Control Procedure Example

Figure 8.2 is a flowchart of the document and data control procedure
and Figure 8.3 is a "Document Change/Release Coversheet" using
the "Owner and User Partnership" system.

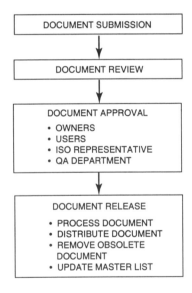

Figure 8.2   Document and Data Control Flowchart

Document Submission

Any employee can request to change and release a document by
submitting a draft and Document Release/Change Form to the
document owner.

```
┌─────────────────────────────────────────────────────────────┐
│              DOCUMENT RELEASE/CHANGE FORM                     │
│                                                               │
│  Document Change Order: _____  Request Date: _____  │
│                                                               │
│  Document Number: _____  Implementation Date: _____   │
│                                                               │
│  Document Title:    _____ │
│                                                               │
│  Originator: _____    Department: _____ │
│                                                               │
│  Document Owner: _____ │
│                                                               │
│  Reason for Change Request: _____ │
│                                                               │
│  _____│
│                                                               │
│  Description of Change:  _____│
│                                                               │
│  _____│
│                                                               │
│                                                               │
│  Document Owner Approval:  _____│
│                                                               │
│                                                               │
│  User Approvals:                                              │
```

User Dept.	Signature	User Dept.	Signature

ISO Representative Approval: _____

Q.A. Approval: _____

Revision History:

Rev.	DCO #	Date	Change Description

Figure 8.3   Document Release/Change Form

Document Review

This step involves a review between the originator, document owner, and user for consensus.

Document Approval

After censensus is obtained, the document is formally circulated for approval to the document approval board (owner, user, ISO representative, and QA).

Document Release

After approval, the document control processes the document, distributes the document to users, retrieves obsolete documents, and updates the master list.

5 How Do We Develop Quality System Documentation?

Developing quality system documentation is a two step approach.

- Design the quality system

- Document the quality system

5.1 Design the Quality System

The design phase is a key activity in the development of quality system documentation. A properly designed system will require fewer modifications later.

The project team is usually responsible for quality system design and the rules are very similar to those used for designing a product.

A good way of designing a quality system is to map or flowchart all activities starting from the customer inquiry through sales, manufacturing, final inspection, and delivery to the customer. Figure 8.4 is an example of how the flowcharting process works for an ISO 9002 organization.

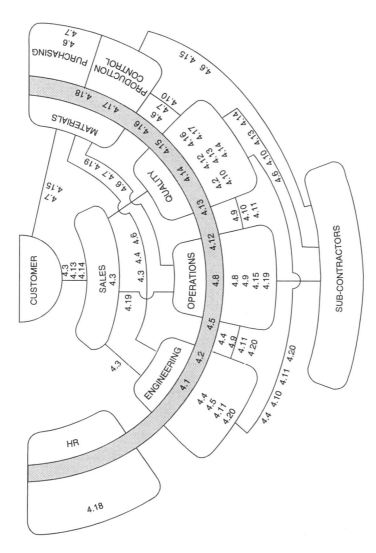

Figure 8.4   System Activity Flowchart

Each department (box) contains requirement numbers the department is responsible for; requirements in the shaded area represent typical company-wide requirements, and the requirements listed on the line indicate the link between the various departments.

Here is a simplified explanation of how the flow chart works.

- **Customer**

  Sales receives customer requirements per 4.3 (Contract Review).

- **Sales**

  Sales reviews the requirements with engineering and materials per 4.3 (Contract Review). If satisfactory, the order is accepted. Note the linkage between sales and engineering.

- **Engineering**

  Engineering provides bill of materials and specifications to materials utilizing 4.3 (Contract Review).

- **Materials**

  Materials purchases components required from sub-contractors per 4.6 (Purchasing) and schedules the order with operations through production control.

- **Quality**

  Quality inspects purchased product per 4.10 (Inspection and Testing) prior to use by operations.

- **Operations**

  Operations produces product conforming to 4.9 (Process Control)

- **Quality**

  Upon completion the quality department inspects product per 4.10 (Inspection and Testing) and if acceptable releases product to materials.

- **Materials**

  Materials ships product to customer per 4.15 (Material Handling, Packaging, Preservation, and Delivery)

5.2 Document the Quality System

Once the quality system has been designed, it must be documented. A documentation "overhaul plan" is shown in Figure 8.5.

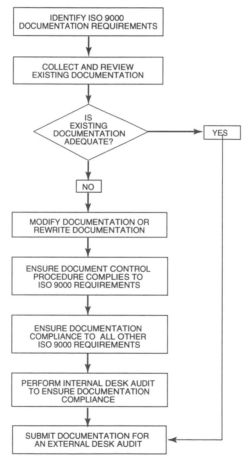

Figure 8.5   Documentation Overhaul Plan

Identify ISO 9000 Documentation Requirements

Every quality system requirement mandates the preparation of documented procedures.

Collect and Review Existing Documentation

Collect all existing quality system documentation and review it for compliance to the applicable standard requirements. A checksheet (Figure 8.6) can be used in this documentation review process.

ISO Requirement	Doc #	Owner	User	Comments
4.1 Management Responsibility	QSD-1A	Top Management	All	
4.2 Quality System	QSD-2B	QA	All	
4.3 Contract Review				No documents available

Figure 8.6   Documentation Review Checksheet

– The International Standard requirements are listed on the checksheet under the "ISO Requirement" column.

– Review each existing document and place the document number in the "Document #" column corresponding to the requirement it meets.

– If not already done, it is wise to determine the proposed "owners" and "users" at this time. This information will speed up the registration effort.

– Any explanations should be written in the "comments" column.

Note: This checksheet is a written record and will be useful to the document owner during the "overhaul" process. This sheet can also be used as a reference during the various audits.

Is Existing Documentation Adequate?

The document review process leads to two conclusions:

Yes – Existing documentation is acceptable

In such a scenario the organization can submit documentation to the registrar for the desk audit.

No – Existing documentation is not acceptable

If the documentation is not acceptable, two options are available:

— Modify documentation

Existing documentation complies partially to the International Standard requirements and minor documentation modifications will bring documentation into compliance.

— Rewrite documentation

Existing documentation is inadequate and needs to be rewritten to comply to the international standard documentation requirements.

Ensure Document Control Procedure Complies to ISO 9000 Requirements

The document control procedure must be compliant with the International Standard requirements first as all other quality system procedures will be issued through it. This procedure should be developed by the project team; however, the ownership to maintain the document control system can be assigned to any department. In most companies document control is associated with the quality or engineering departments.

Note: It is possible to have multiple document control procedures; for: policies, procedures, process work instructions, engineering changes/tests, drawing control, etc., if such an arrangement makes sense to the organization.

Ensure Documentation Compliance to Other ISO 9000 Requirements

Company-Wide Requirements

These requirements are used by everyone in the company and should be documented first. Listed below are general company-wide requirements:

- Management responsibility
- Quality system
- Document and data control
- Product identification and traceability
- Inspection and test status
- Control of non-conforming product
- Corrective and preventive action
- Handling, storage, packaging, preservation, and delivery
- Control of quality records
- Internal quality audits
- Training

Area-Specific Requirements

These are requirements not used by everyone in the company. Listed below are area specific requirements:

- Contract review
- Design control
- Purchasing

- Control of customer-supplied products

- Process control

- Inspection and testing

- Control of inspection, measuring, and test equipment

- Servicing

- Statistical techniques

Perform Internal Desk Audit to Ensure Documentation Compliance

Upon the completion of quality system documentation the project team conducts an internal desk audit to ensure that compliance to the requirements of the International Standard has been achieved.

Any issues or non-conformances are resolved at this time.

Submit Documentation for External Desk Audit

When the project team is satisfied with the quality system documentation, the appropriate documentation is submitted to the registrar for an external desk audit.

5.3 How Do We Implement Quality System Documentation?

If designing the QSD was: "document what you do," then implementing QSD is: "doing what you document." Believe it or not implementation can be fun! "Building the airplane is not half the fun of flying the airplane."

Implementation also has to be carefully planned. Implement the document and data control documentation first. For other documentation there is no one best plan. Some companies choose to implement individual procedures upon their completion, some implement all together, others start at the inputs (contract review, purchasing) to the system and work their way into the system in a logical progression.

There should be no surprises since every owner and user is thoroughly familiar with the documentation. If there are some problems, don't panic, that's where continuous improvement comes in.

It may be a good idea to perform a trial implementation on a limited scale. Remember that the organization will not be ready for either a "desk audit" or a "pre-registration audit" without complete documentation of the quality system. Also remember that it is not sufficient to have the documentation completed, these documents should be seasoned by use. Some documentation such as corrective action must be in use long enough to have collected a history of corrective actions through incident to closure.

6   Conclusion

As each one of the requirements of the International Standard becomes understood, the components of your quality system, together with the appropriate documentation will, like bricks in a structure, fall into place.

It is important to remember that development and implementation of QSD is not an event, but an ongoing process and is to be continually improved using TAP-PDSA. Always keep in focus that the primary purpose of a quality system is to ensure customer satisfaction and profitability, not just ISO 9000 certification.

Chapter 9
Life After Registration

If the organization is now ISO 9000 registered, Congratulations! Hopefully everybody has had time to enjoy the success.

After registration, the company has an opportunity to look back at what has been accomplished and to look forward to what the future holds.

- Was ISO 9000 registration worth the effort?

- What's next?

1 Was ISO 9000 Registration Worth the Effort?

To answer this question, it is necessary to go back to the reason why the registration project was initiated in the first place, and to verify that the objectives were met.

Note: This is the study step in the TAP-PDSA.

157

If the reason was:

- **Continuous Improvement**

  Are the yields increasing? Are the reject ratios declining? Are on-time requirements being met?

- **Marketing Strategy**

  Was there an increase in sales? Has the market share gone up? Did the company acquire new customers as a result of registration?

- **Customer Requirements**

  Are the customers satisfied?   Are there signs of improved customer relations? Did the company receive any customer awards?

If the answers to the above questions were less than an unqualified yes, then the company needs to continue the TAP-PDSA cycles to meet the objectives by taking the appropriate corrective actions.

2   What's Next?

ISO 9000 registration is only a minimum requirement.   The company should now plan to take the quality system to the "next level" as explained by the following:

- Sustain registration through continuous improvement
- Look at specific customer requirements
- Evaluate the global impact on the quality system
- Prepare for future generation requirements

2.1 Sustain Registration Through Continuous Improvement

After the company is registered, it must sustain registration through a system of periodic surveillance audits. Experience has shown that many companies "take it easy" after becoming registered and the first surveillance audit comes as a rude awakening that the system is not performing as effectively as it should.

This echoes Deming's first point of management: "Constancy of purpose for continuous improvement." Maintain ISO 9000 registration diligently and build upon it. ISO 9000 registration is an excellent platform to scale the heights of quality excellence, not a signal that the quality project has been completed.

Constantly improve the system (4 M's and E) to improve quality and delivery, reduce costs to stay ahead of competition.

Use TAP-PDSA to re-evaluate the goals and objectives of the system. Develop short-term and long-term goals. For example, select a few quality system requirements each quarter and take them to the next level of improvement as shown in Table 9.1.

Table 9.1: Quality System Improvement

Quality System Requirement	Next Level Development
Management Responsibility	Management Measurement System
Quality System	Simplify System Structure
Document and Data Control	Paperless Document Control System
Purchasing	Supplier Management Team
Product Identification and Traceability	Bar Code System
Quality Records	Paperless Record Keeping System
Internal Quality Audits	Cross Functional Audit Teams
Training	Internal Training University
Statistical Techniques	Company-Wide SPC System

Example: Develop a supplier management team (materials, quality, engineering, operations) and drive quality requirements down to the sub-contractors. Move the quality system requirements onto the supplier base. "Ship to stock" has the potential for immense cost savings and quality improvements. Use the internal auditors to evaluate and assist suppliers to become ISO 9000 compliant.

Similarly, select other quality system requirements over a period of time and improve them. This overall plan becomes a **strategic quality improvement plan (SQP)**.

2.2 Look at Specific Customer Quality Requirements

ISO 9000 is a baseline standard. Many industries have adopted ISO 9000 or customized it to develop their specific quality system requirements such as QS-9000 for the automotive industry. Identify the customer's changing requirements and work towards complying to them. Some customer-specific quality system requirements are:

*   QS-9000

*   TE-9000

*   DoD

2.2.1 QS-9000

What Is QS-9000?

QS-9000 is the quality system standard of Ford, Chrysler, General Motors, truck manufacturers, and other subscribing companies. It defines their quality system expectations.

Who Does QS-9000 Apply To?

QS-9000 applies to all internal and external suppliers of: a) production materials, b) production or service parts, c) heat treating, painting, plating, or other finishing services directly to Chrysler,

Ford, General Motors, or other OEM customers subscribing to QS-9000

What Are Deadlines for QS-9000 Registration?

- Chrysler requires tier one QS-9000 registration by **July 3, 1997**

- General Motors requires tier one QS-9000 registration by **December 31, 1997**

- Ford does not require 3rd party registration at this time

Note: **Tier one suppliers** are suppliers that do business with the big three directly.

Why Was QS-9000 Developed?

QS-9000 was developed by the big three supplier quality group to harmonize their quality requirements (Table 9.2).

Table 9.2: Harmonization of Quality System Requirements

Company	Before	Today
Ford	Q-101	QS-9000
Chrysler	SQA	QS-9000
GM	Targets for Excellence	QS-9000
Truck Mfg.	N/A	QS-9000

What Is the Structure of the QS-9000 Standards?

The QS-9000 standard is organized in the following three sections:

Section I: ISO 9000-Based Requirements

Section I is known as the ISO 9000-based requirement and contains the ISO 9001 requirements in their entirety. Also included in this section are additional requirements of the big three. The ISO 9001 requirements are in italic while additional requirements are in roman type.

Section II: Sector-Specific Requirements

This section contains sector-specific requirements and addresses the following three topics:

*   Production Parts Approval Process

*   Continuous Improvement

*   Manufacturing Capabilities

Section III: Customer-Specific Requirements

This section addresses the customer-specific requirements for Chrysler, Ford, General Motors, and truck manufacturers, such as third party registration requirements, identification symbols, etc.

The overall QS-9000 structure is shown in Table 9.3.

Table 9.3:   QS-9000 Structure

Section I ISO 9000-Based Requirements	Section II Sector-Specific Requirements	Section III Customer-Specific Requirements
ISO 9001 + Big Three Requirements	1) Production Part Approval Process 2) Continuous Improvement 3) Manufacturing Capabilities	1) Chrysler 2) Ford 3) General Motors 4) Truck Manufacturers
Reference Manuals • Advanced Product Quality Planning and Control Plan (APQP) • Failure Mode and Effect Analysis (FMEA) • Measurement System Analysis (MSA) • Fundamental SPC	**Reference Manuals** • Production Part Approval Process (PPAP)	

Who Else Has Adopted QS-9000?

QS-9000 is fast becoming the de facto automotive quality system standard and to date has been adopted by the following:

- Toyota
- Mazda

2.2.2 TE-9000

What Is TE-9000?

TE-9000 is the quality system standard for suppliers of automotive tooling and equipment to the big three. The QS-9000 standard pertains to the production and service parts that are utilized in the vehicle while TE-9000 requirements apply to the tooling and equipment that are used to build a vehicle.

Who Will TE-9000 Apply To?

Suppliers whose processes involve assembly, balancing, casting, forging, forming, gauging, heat treating, machining, material handling, measuring, molding, packaging, painting, plating, stamping, tooling, and welding, would be prime candidates for TE-9000 compliance.

2.2.3 DoD

The US military is actively replacing military specifications with commercial standards.

Mil-Q-9858A, the defense quality system standard, was inactivated in August 1995. The defense department has designated ISO 9001 and ISO 9002 as the quality system standards of choice for its suppliers. The DoD will require ISO 9000 compliance but not registration.

Note: It is interesting to note that Mil-Q-9858A and Mil-I-45208 were utilized as a major source of input during the development of the ISO 9000 series of standards.

2.3   Evaluate the Global Impact on the Quality System

Prepare the quality system diligently for a global economy. Surveys indicate that 75% of companies that do not market globally will do so by the end of the decade or risk going out of business. Don't be late. Those companies planning to market products globally must be aware of the European Union (EU) and the product registration mark CE.

2.3.1 European Union

What Is the European Union? and What Does It Do?

The European Union (EU) is a treaty organization. It originated with the 1957 Treaty of Rome. One of its objective was to abolish tariffs and trade quotas among the member countries and to stimulate economic growth within the member states. The EU acts as a sovereign legal entity controlling the activities assigned to it by the treaty. An early target for the EU was defining national product certification requirements. Differences in product certification requirements made selling products in the various countries a difficult undertaking because of the need for duplication of conformity tests, certification, documentation, and the requirement for separate approvals, often from both national and local regulatory agencies.

The Treaty of Rome was amended in 1986 to incorporate the "Single European Act", which established the principle that products that meet the requirements of one EU member state could freely circulate in other member states. This is a concept which is very similar to the Interstate Commerce article in the US constitution. All of this single market activity became known as EC 92 and was scheduled for implementation at the end of the 1992 calendar year. The target year was found to be unrealistic and the goal is presently being implemented progressively.

EU and ISO

Just as ISO was founded in 1946 with a goal of commonly accepted standards to allow countries to better use each other's commodities, manufactures, and products, the EU has expanded this concept through EU directives to harmonize standards that provide conformity assessment procedures that are consistent throughout the 15 nation union and to provide regulations for competent certification and testing bodies.

However, the differences are immense. ISO is a voluntary standards organization and the EU is a legislating force. Its rules are binding among member countries and any company that wishes to market a product in the EU. Individual companies cannot work directly with the EU. This must be accomplished through Mutual Recognition Agreements (MRA) negotiated with non-EU countries.

EU and USA

A key statement was made by a European Commission spokesperson at the December 1995 meeting in Madrid, Spain, convened for the purpose of signing a major trade and economic cooperation accord with the United States of America: "We wish to put particular emphasis on the pledge made by the two trading partners (United States and the European Union) to work towards mutual recognition of one another's product standards and certification. Europe and America will become more 'open' for business. If a product is made in Europe, it must be good enough for America and if made in America, it must be good enough for Europe."

2.3.2 CE Mark

Quality systems must become increasingly sophisticated to accommodate the more stringent product quality requirements that are coming out of the European Union, and North America (particularly the United States and Canada). Go into any electronics store and inspect the back of the base on a piece of equipment. On many pieces you will find these approval labels: UL, CSA, TUV/ GS, BZT, and CE mark.

The CE mark is the most progressive.

What Is a CE Mark?

The CE mark is a product registration mark which identifies that the product was manufactured under EU normalization directives. The European Union (EU) has adopted a number of directives that require certain products entering it's market to bear a CE mark and appropriate technical documentation indicating compliance to the requirements of the directive.

The following are some directives:

89/336/EEC - The Machinery Directive. This directive covers all machinery, defined as "an assembly of linked parts or components, at least one of which moves, with the appropiate actuators, control and power circuits etc., joined together for a specific application, in particular for the processing, treatment, moving or packaging of a material." The Directive outlines the essential safety requirements for this equipment. It was adopted on Jume 14, 1989 and went into effect on January 1, 1995.

93/42/EEC - The Medical Devices Directive. This directive covers medical devices and their accessories. For the purposes of this direcive, accessories shall be treated as medical devices in their own right. Definations of "medical devices: can be found in the article 1 Definations, scope 2 (a) of this Directive. This directive outlines the scope, classifications, conformity assessment procedures and essential requirements. It was adopted on June 14, 1993 and goes into effect on June 14, 1998.

Products under the scope of directives will not be allowed to be sold in the European Market unless they bear the appropriate CE mark and technical documentation.

What Is the Purpose of the CE Mark?

The purpose of the CE Mark is to provide for the "free movement" of goods across national boundaries. The EU established legal requirements known as directives, which are intended to harmonize the existing legal practices in the member states. The national governments of the member states are obligated to harmonize national law with the directives. Any national laws that conflict with the directives are required to be repealed.

How Does an Organization Obtain a CE Mark for Its Product?

The CE mark can be obtained through self declaration or through the use of a notified body:

• Self Declaration (declaration of conformity)

The manufacturer performs tests to the appropriate EU directive and prepares a test report indicating that the product complies with the appropriate directive. Upon compliance, the manufacturer affixes a CE mark to the product.

• Use Of a Notified Body

For certain products, a notified body must be used to perform or verify the required tests and upon compliance provide a certificate of conformity. The manufacturer is then allowed to affix the product with a CE mark.

Notified Body: A notified body is an agency considered competent in the field of a specific directive and who is authorized by EU member states to test, verify and certify product per the directive.

Note: Some EU directives may require ISO 9000 registration as a condition to obtaining the CE mark.

2.4 Prepare for Future Generation Requirements

There are several standards under development today which will become the requirements of tomorrow. To stay ahead, evaluate their impact on the company's business plan. The following standards will be of significance in the next few years.

- ISO 14000, Environmental Management System Standards (EMS)
- Future ISO 9000 Stanadards (1999)

2.4.1 ISO 14000

What Are the ISO 14000 Series of Standards?

In 1993, ISO created TC-207 to develop environmental standards, i.e., the ISO 14000 Series of Standards.

Table 9.4 lists the ISO 14000 series of standards under development.

Table 9.4:   ISO 14000 Series of Standards

ISO 14001:	Environmental Management System Specification (analogous to ISO 9001)
ISO 14004:	Environmental Management System–General Guidelines on Principles, Systems, and Supporting Techniques
ISO 14010:	Guidelines for Environmental Auditing–General Principles of Environmental Auditing
ISO 14011:	Guidelines for Environmental Auditing–Audit Procedures-Part I: Auditing of Environmental Management Systems
ISO 14012:	Guidelines for Environmental Auditing–Qualification Criteria for Environmental Auditors

Note: Standards pertaining to the environmental labeling, environmental performance, life cycle assessment, and terms and definitions are expected to emerge.

What Is the Purpose of ISO 14001?

ISO 14001 specifies the requirements of an environmental management system.

Compliance to ISO 14001 will demonstrate an organization's ability to meet environmental objectives and targets. ISO 14001 is generic in nature, and it will therefore, be applicable to any type of organization.

2.4.2 The Future ISO 9000 Standards (1999)

The ISO 9000 series of standards were revised in 1994. The next major revisions are expected in the year 1999. Based on the outcome of the TC 176 annual meeting in Durban, South Africa, the following are insights into the future of ISO 9000 and its series of standards:

ISO 9001 Consolidation

ISO 9001, 9002, and 9003 may be consolidated into a single standard ISO 9001 to be published around the year 1999. Companies currently registered to ISO 9002 will then be registered to ISO 9001 without design control, but for companies currently registered to ISO 9003, the upgrade is expected to involve additional requirements.

ISO 9001 Reorganization

Speakers at the ISO 9000 forum symposium looked at the future architecture of ISO 9001. A concept plan showed that the twenty clauses had been consolidated into the following major groupings:

- Executive Management
- Process Management
- Measurement
- Evaluation and Improvement

Note: The focus of the changes is primarily on customer's requirements and measurements.

ISO 9000 Consolidation

ISO 9000 document may consolidate 9000-1,2 etc., and become a more comprehensive document. ISO 9000 is also projected to include an expanded vocabulary and technical terms, presently ISO 8402.

3   Conclusion

Changes to the standard are going to be more dramatic in the next 5 years than they have been in the last 25 years. The global market is forcing the globalization of standards. Competition has become fiercer and to stay in business companies are required to continually comply with the stringent emerging requirements.

Train and educate continuously to learn about future trends and requirements, then perform an **analysis** of the existing situation to get a better understanding, and finally predict a **plan** with the highest probability of success. You will then "**lead the way.**"

Bibliography

Beaumont, Leland R. *ISO 9001, The Standard Companion* (2nd ed). Middletown, NJ: ISO Easy. 1993

Fellers, Gary. *The Deming Vision: TQM for Administrators.* Milwaukee, WI: ASQC Quality Press, 1992

Clements, Richard Barrett. *Quality Manager's Complete Guide to ISO 9000.* Englewood Cliffs, NJ: Prentice Hall, 1993

Crosby, Phillip B. *Quality Is Free, The Art of Making Certain.* New York, NY: Mc Graw Hill, 1979

Delavigne, Kenneth T. and Robertson, J. Daniel. *Deming's Profound Changes:* Englewood Cliffs, NJ: PTR Prentice Hall, 1994

Deming, W. Edwards. *Out Of Crisis.* Cambridge, MA: MIT Center for Advanced Engineering Study, 1986

171

Deming, W. Edwards. *The New Economics for Industry, Government and Education.* Cambridge, MA: MIT Center for Advanced Engineering Study, 1993

Gluckman, Perry and Roome, Diana Reynolds. *Everyday Heroes, from Taylor to Deming; The Journey to Higher Productivity.* Knoxville, TN: SPC Publications, 1990

International Organization for Standardization. *ISO 9000 International Standards for Quality Management.* Geneva, Switzerland, ISO, 1994

Ishakawa, Kaoru. *Guide to Quality Control* (2nd ed.). Tokyo, Japan: Asian Productivity Organization (in the United States, UNIPUB, New York, NY) 1985

Johnson, L. Marvin. *Quality Assurance Program Evaluation* (revised edition). Los Angeles, CA: L. Marvin Jones, Associates, 1990

Juran, J.M. *Management Breakthrough.* New York, NY: Mc Graw Hill, 1964

Kanter, Rob. *ISO 9000 Answer Book.* Essex Junction, VT: Oliver Wright Publications, 1994

Kivenko, Kenneth. *Quality Control for Management.* Englewood Cliffs, NJ: Prentice Hall, 1984

Lamprecht, James L. *ISO 9000: Preparing for Registration.* Milwaukee, WI: ASQC Quality Press, 1992

Lamprecht, James L. *Implementing the ISO 9000 Series.* New York: Marcel Dekker and Milwaukee, WI: ASQC Quality Press, 1993

MacLean, Gary E. *Documenting Quality for ISO 9000 and Other Industry Standards.* Milwaukee, WI: ASQC Quality Press, 1993

Mills, Charles A. *The Quality Audit, A Management Education Tool.* New York, New York, NY: McGraw Hill, 1989

Peach, Robert W. (editor). *The ISO 9000 Handbook* (2nd ed.). Fairfax, VA: CEEM Information Services, 1994

Sayle, Allan J. *Meeting ISO 9000 in a TQM World* (2nd ed.). London, England (UK): AJSL, 1994

Senge, Peter. *The Fifth Discipline, The Art and Practice of the Learning Organization.* New York, NY: Doubleday/Currency, 1990

Scherkenbach, William W. *The Deming Route to Quality and Productivity: Roadmaps and Roadblocks.* Washington, DC: CEEP Press Books, 1986

Walton, Mary. *The Deming Management Method,* New York, NY: The Putnam Publishing Group (Petigree), 1986

Index